MATHEMATICAL MODELS OF MEANING

MATHEMATICAL MODELS OF MEANING

A Dynamic Systems Approach to Possible World Semiotics

Paul Kockelman

The MIT Press
Cambridge, Massachusetts
London, England

The MIT Press
Massachusetts Institute of Technology
77 Massachusetts Avenue, Cambridge, MA 02139
mitpress.mit.edu

This book was set in Times New Roman by Paul Kockelman. Printed and bound in the United States of America.

Library of Congress Cataloging-in-Publication Data

Names: Kockelman, Paul, author.
Title: Mathematical models of meaning : a dynamic systems approach to possible world semiotics / Paul Kockelman.
Description: Cambridge, Massachusetts : The MIT Press, 2025. | Includes bibliographical references and index.
Identifiers: LCCN 2024038530 (print) | LCCN 2024038531 (ebook) | ISBN 9780262552684 (paperback) | ISBN 9780262383486 (pdf) | ISBN 9780262383493 (epub)
Subjects: LCSH: Semiotics. | Mathematical linguistics. | System theory.
Classification: LCC P99.4.M63 K63 2025 (print) | LCC P99.4.M63 (ebook) | DDC 410.1/51–dc23/eng/20241214
LC record available at https://lccn.loc.gov/2024038530
LC ebook record available at https://lccn.loc.gov/2024038531

10 9 8 7 6 5 4 3 2 1

EU product safety and compliance information contact is: mitp-eu-gpsr@mit.edu

For Mia, Zeno, and Lara

Contents

List of Figures

Preface

This book offers a mathematical model of meaning, and thereby provides answers to the following kinds of questions: What is meaning? What is the relation between meaning, information, value, and purpose? What ingredients are necessary for a system to exhibit meaning? What behaviors, and capacities for behavior, are particular to meaning-oriented agents? Is there a relatively simple mathematical model that can adequately capture the dynamics—and diversity—of meaning-oriented agents? How do we best bridge the divide between interpretive paradigms that are qualitative and context-rich and formal methods that are quantitative and domain general?

At the center of this model is a distributed agent that can sense and instigate relatively immediate events and, through these, project and effect relatively mediate events, in reference to a dynamic set of commitments and values (understood as an interpretive ground), and by means of a double integration over past and future worlds.

This book argues that interpretive grounds are central to meaningful processes. It argues that such grounds can be embedded in environments no less than enminded in organisms, and hence turn on relatively objective patterns and resources no less than relatively subjective commitments and values. And it shows that such grounds function as dynamic variables: at once shaped by meaningful processes and shaping of meaningful processes. As will be seen, such a dynamic coupling between figures and grounds, qua interactional practices and interpretive resources, makes this mathematical model of meaning particularly rich and revealing.

In offering such an analysis, this book brings together the objects of signs and the ends of agents, and hence motivation as much as meaning. It connects agents that can select (insofar as they can choose different courses of action in real time) and agents that are selected (such that they can evolve over generational time). It accounts for the behavior of agents that are oriented to diverse kinds of value: from expected utility to free energy, from biological fitness to social status. It shows the connection between the possible worlds of formal semantics and the microstates of statistical physics. And it puts the fixed points of dynamic systems theory into relation with the hermeneutic circles of critical social theory.

While the model incorporates core ideas from a pragmatist tradition, it weaves together a range of powerful ideas from other paradigms, including Bayesian inference, statistical mechanics, decision theory, mathematical biology, evolutionary game theory, possible world semantics, machine learning, linguistics, and anthropology. Its analytic framework thereby provides a relatively seamless integration of distinct methods and theories.

After introducing the model, and reviewing its core assumptions, chapters 2 through 9 explore the entailments of the model, and assess its merits, by using it to analyze a variety of increasingly complex scenarios. As will be seen, the math is done in a complete, but conversational way. And the formalism begins simply and ramps up slowly, such that a wide range of readers will be able to understand the concepts, follow the arguments, imagine novel scenarios, and extend the analysis themselves.

Acknowledgments

Greg Urban and Liam Taylor read a draft of this manuscript and gave me incredibly helpful suggestions. Philip Laughlin, my editor at MIT Press, was also very supportive. Jitendra Kumar, the senior project manager, was extremely helpful and patient. Thank you.

A range of friends and interlocutors transformed my thinking about various parts of this argument. Many thanks to members of the Semiotics Working Group at the Center for Transcultural Studies, especially Benjamin Lee, Terra Edwards, Andrew Carruthers, Katie Stewart, Miyako Inoue, Bill Hanks, Greg Urban, Xochitl Marsilli-Vargas, and Kamala Russell. I also found conversations with Siri Lamoureaux, Robert Meister, Mike Cepek, Nick Enfield, Gary Tomlinson, Jonathan Beller, and Julia Elyachar to be very stimulating.

Elijan Mastnak radically altered my relation to Arch Linux and LaTeX.

Support for this project was provided by the MacMillan Center for International and Area Studies at Yale University. Time for writing was provided by a generous research leave from Yale University. Many thanks to the Department of Anthropology, with a loud shout out to Doug Rogers, Erik Harms, and Marleen Cullen.

Special thanks to Erik Thogersen, Jessica Strick, Zoe Zoe Zoe, and everyone that Grannie Valerie brought into being.

1 Introduction

Overview

Section 1.1 summarizes some relevant ideas from Charles Sanders Peirce's theory of meaning, and highlights the ways that this book both builds on them and goes beyond them. It describes the core components of this model of meaning in a non-mathematical way. Section 1.2 projects a mathematical framework onto these components. It motivates and explicates the main equation of the model and highlights some of its key assumptions. Section 1.3 shows how the main equation can be applied to agents capable of evolving on phylogenetic time scales, as well as agents capable of learning on interactional time scales. Section 1.4 focuses on signaling processes and the division of semiotic labor. Section 1.5 distinguishes between the meaning, value, and information of a sign, and offers a characteristic measure for each of these three properties. Section 1.6 summarizes the contents and arguments of the chapters that follow, and section 1.7 surveys the relevant literature.

1.1 The Core Components

Charles Sanders Peirce (1839–1914) was an American philosopher and logician, famous not only for his contributions to logic, mathematics, and meaning, but also for being the foundational theorist of pragmatism. In the rest of this section, we describe this model of meaning in relatively qualitative terms, focusing on the ways it resonates with—but also routes around—some of Peirce's core assumptions.

In a Peircean view of meaning, a *semiotic process* involves three interrelated components: a *sign* is whatever stands for something else; an *object* is whatever is stood for by a sign; and an *interpretant* is whatever a sign creates, insofar as it is taken to stand for an object. For example, a father points (sign) to a bird (object) and his daughter turns to look (interpretant). A gold or copper coin is pulled from an urn (sign), indicating a particular composition of coins within the urn (object), and the urn is then accepted or rejected (interpretant). Finally, a student raises their hand (sign), indicating their desire to ask a question (object), and the teacher calls on them (interpretant).

Unlike many other accounts of meaning, which foreground a single relation of standing for (e.g., signifier-signified, expression-referent, signal-message, and so forth), Peirce's account foregrounds a relation between two such relations. Loosely speaking, *a sign stands for its object on the one hand, and its interpretant on the other, in such a way as to make the interpretant stand in relation to the object corresponding to its own relation to the object.* These three components, as well as this relation between relations, are shown in figure 1.1.

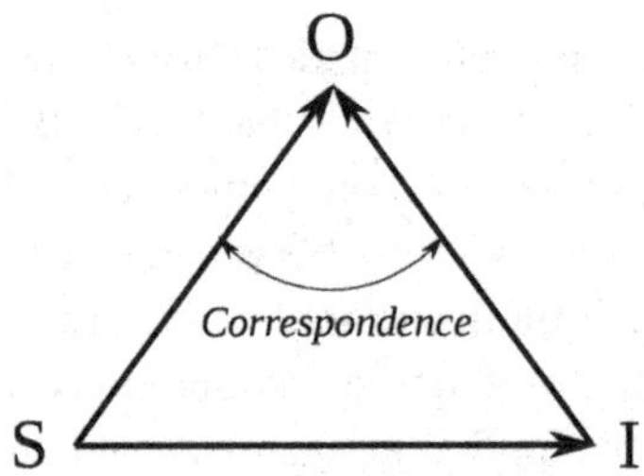

Figure 1.1: Signs, Objects, and Interpretants

While there is some slippage in Peirce's work between the interpretant as an agent (or 'mind') that interprets the sign, and the interpretant as another sign (brought into being by the agent's interpretation of the first sign), we will be careful to distinguish between the two senses. In the scenario just sketched, for example, the father is the signifying agent, or *signer*, and his pointing gesture is the sign; the daughter, in contrast, is the interpreting agent, or *interpreter*, and her turning to look is the *interpretant*. The sign-interpretant relation is thus both mediated by, and mediating of, this signer-interpreter relation. Phrased another way, a relation between two entities (the sign and the interpretant) mediates a relation between two agents (the signer and the interpreter). As will be seen, such social relations—and relations between relations more generally—are often the impetus for and outcome of semiotic processes.

In contrast to Peirce, who tended to focus on *significance* (or meaning, in the sense of 'standing for'), we will also focus on *selection* (qua motivation, or meaning in the sense of 'striving for'). This model therefore includes not only the objects that signs stand for (and the interpretants that signs create), but also the values that agents strive for and/or the functions that agents serve.

Phrased another way, semiosis, and hence both signification and interpretation, are mediated not just by teleological, teleonomic, and teleomatic processes (broadly construed), but also by 'teleognomic' processes (Enfield and Kockelman 2017). Therefore, while only a few such agents are capable of selecting (in the sense of self-consciously choosing, as stereotypically understood), almost all such agents are capable of being selected (whether via natural selection, reinforcement learning, algorithmic sieving, design, manufacture, discipline, regimentation, framing, and the like).

In short, semiotic agents are not just the instigators, but also the outcomes, of various modes of selection. And so aside from their shared capacity to signify and interpret, such agents are radically heterogeneous. As will be seen, this model of meaning can be used to analyze the behavior of a wide range of agents, some of which are self-conscious, well-informed, technologically sophisticated, and far-seeing, and some of which are at the limit of what counts as life.

In contrast to Peirce, who foregrounded three components in his analysis of semiotic processes (sign, object, and interpretant), we will sometimes include not just the agent, but also a fourth component: the *consequent*. Just as we distinguish between the sign (as whatever is sensed by the agent) and the object (as whatever is mediated by the sign), we also distinguish between the interpretant (as whatever is instigated by the agent) and the consequent (as whatever is mediated by the interpretant). In the first scenario to be examined, the sign will be a symptom, the object will be an illness, the interpretant will be the action of taking (or not taking) a medication, and the consequent will be the effect, and/or side effect, that the medication has when taken. While both kinds of entities (objects and consequents) are 'absent' (in the sense of not being directly sensed or instigated by the agent), they are absolutely 'present' (in the sense of being indirectly sensed and instigated by the agent through such mediating signs and interpretants, whatever the degree of remove).

Phrased another way, just as there is a kind of *experiential slash* between the sign (which the agent can directly sense) and what it stands for (the object), there is also a kind of *agentive slash* between the interpretant (which the agent can directly instigate) and what it results in (the consequent).

All the foregoing components may be put into relation, and thereby diagrammed and described, as follows: *The relation between what the agent (A) instigates (I) and what it senses (S) makes sense in relation to the relation between what the agent effects (C) and what it projects (O), given the relation between that relation (O/C) and the agent's identity, origins, and/or interests.* See figure 1.2. A key argument of this book is that meaning resides in such interrelations.

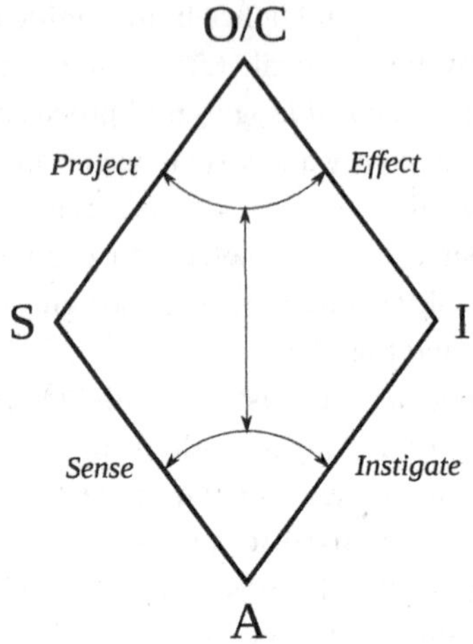

Figure 1.2: Relations between Relations

As will be seen, where we—as analysts—draw the boundary between object and sign or interpretant and consequent, and hence how we determine where sensation ends and projection begins, or where instigation ends and effecting begins, is often a frame-dependent and/or scale-specific decision (Kockelman 2011). For example, and more generally, the interpretant of a prior semiotic process may constitute the sign in a subsequent semiotic process; the sign of a lower-order semiotic process may constitute the object of a higher-order semiotic process; the difference between interpretants and consequents, or between objects and signs, may sometimes be collapsed; affects, inferences, outputs, and traits may be framed as interpretants no less than actions, habits, utterances, and responses; a single, relatively distributed agent may incorporate several smaller agents; most selecting agents were themselves selected; agents that select often employ agents that were selected; semiosis may mediate relatively private cognitive processes no less than relatively public communicative practices; thinking, evolving, and learning are irreducibly coupled; and so on, perhaps indefinitely.

While Peirce—and certainly Peirceans—tend to focus on sign-object relations through the frame of his most famous trichotomy (icons, indices, and symbols), not much will be made of these categories here. Instead, we will focus on the sensibilities and assumptions that agents have regarding not only sign-object relations, but also interpretant-object and consequent-interpretant relations. When dealing with human-like agents, such sensibilities and assumptions will often be framed as propensities and probabilities, and hence as intensities of desire (or affect more generally) and degrees of commitment (belief, doubt, and/or uncertainty). Such sensibilities and assumptions may be referred to as the *grounds*, or 'guiding principles', of semiotic processes. As will be seen, not only can such interpretive grounds constitute the objects of meta-semiotic processes (and thereby become 'figures'), they can also be transformed by semiotic processes. They thereby constitute a relatively fluid, as opposed to a fixed, semiotic resource.

Such grounds might thereby be likened to the metric in general relativity, insofar as they constitute dynamic variables that enable agents to measure quantities, or at least register intensities, in ways that take into account invariance no less than variation. And, even more crucially, just as the metric is both determined by the distribution of matter and determining of the flow of matter, interpretive grounds are at once shaped by semiotic processes and shaping of semiotic processes. As will be seen, and as might be expected, such a coupling can lead to somewhat complicated equations of semiosis, with rich and revealing solutions.

In short, it is assumed that (humanlike) semiotic agents can engage in relatively immediate processes of sensation and instigation, meaning that they can take differences as inputs (and thereby be affected) and make differences as outputs (and thereby be effective), where a difference is some kind of differentiable event. It is also assumed that agents can engage in relatively mediate processes, such that they are sensitive to whatever causes their sensations and whatever is caused by their instigations. Not only can such agents relate to the mediate through the immediate (in two directions), they also have a model of such relations (however embodied or engenomed, erroneous or simplistic, innate or unconscious). In particular, they have a sense of the organization and intensity with which the immediate and mediate are coupled: what kinds of objects lead to what kinds of signs, and what kinds of consequents follow from what kinds of interpretants, and with what degree of strength, likelihood, or frequency. Finally, they have a sense—however unconscious, instinctual, misguided, or limited—of the relative value of various relations among such events; and hence a sense of which worlds they would rather reside in, and well as which actions they should undertake, in which contexts, to better realize such worlds.

Crucially, saying that such agents have a 'sense' of such interrelations does not mean that they must consciously register them in any way. It only means that analysts must take into account such relations if they are to make sense of the behavior, identity, and/or origins of such agents. Such a strong, and perhaps strange claim will become clearer in the sections that follow.

While Peirce's account of semiosis is frequently used to conduct relatively qualitative studies of real-time semiotic processes as they unfold in 'the real world' (by anthropologists, linguists, historians, discourse analysts, critical theorists, and like-minded scholars), we use it—or at least our extension and transformation of it—to offer a mathematical model of meaning. Peirce himself, of course, made great contributions to logic, statistics, and inference; and so he would surely be comfortable with the quantitative and logical direction that his usually qualitatively deployed ideas will be taken. (However trenchant his critique of the limits of this particular approach might have been.) The strategy used in this book, however, is relational as opposed to quantitative. Rather than crunching numbers, proving theorems, or running computer simulations, we will elucidate the *critical points* of relatively abstract scenarios, which are chosen to foreground key aspects of the model of meaning being offered in relation to classic topics from a variety of literatures. In this way, we usually strive to find analytic solutions to illuminating scenarios, along with their fixed points, singularities, and limits.

Moreover, this book is not just meant to offer a mathematical model of meaning; it is also meant to put quantitative approaches to meaning in dialogue with qualitative approaches, as well as to put naturalist approaches in dialogue with critical and hermeneutic approaches. This is not just done as a bridge-building enterprise, such that scholars and scientists of different persuasions can be in conversation. It is also done to offer qualitative theorists (in history, anthropology, media studies, critical theory, and the like) a framework for analyzing the history and practice of mathematical modes of meaning, many of which constitute key agents and intermediaries in the infrastructure of modern life.

The next section lays out some of the key features of this mathematical framing in a more formal way, and with a focus on relatively rational agents. As will be seen, it assumes that readers have a basic understanding of the mathematics of probability, expected value, and related notions. Readers desiring a relatively simple and concrete scenario, with each mathematical term explained in detail and each calculation carried out in full can look to chapter 2, which provides a relatively easy transition to some of these technicalities.

1.2 The Main Equation

Building on the foregoing conceptual model, an *interpretive ground* may be more formally represented as an eight-tuple $\langle S, I, O, C, P, L, E, V \rangle$. The first four elements in this ordered list constitute an *ontology*, understood as the following four core variables (along with their possible values):

- A set of *signs*, $S = \{S_1, S_2, S_3, \ldots\}$, understood as whatever an agent can sense or register more generally (where $|S|$ will denote the number of signs in this set, and similarly for the other three variables that follow);
- A set of *interpretants*, $I = \{I_1, I_2, I_3, \ldots\}$, understood as whatever an agent can instigate, update, or otherwise directly do;
- A set of *objects*, $O = \{O_1, O_2, O_3, \ldots\}$, understood as whatever an agent can project, infer, learn about, or come to know, given what it senses;
- A set of *consequents*, $C = \{C_1, C_2, C_3, \ldots\}$, understood as whatever an agent can effect, produce, accomplish, modify, maintain, or otherwise indirectly bring about given what it instigates.

The next three elements in the tuple constitute *commitments*, which may be modeled as conditional probability distributions (along with a presupposed causal structure) as follows:

- The *priors*, $P(O_i)$, where index i ranges over the objects in the ontology such that the priors may usually be represented as an $|O| \times 1$ matrix;
- The *likelihoods*, $P(S_i/O_j)$, where the indices range over the signs and objects in the ontology such that the likelihoods may usually be represented as an $|S| \times |O|$ matrix;
- The *eventualities*, $P(C_i/I_j \wedge O_k)$, where the indices may range over the consequents, interpretants, and objects in the ontology such that the eventualities may usually be represented as a $|C| \times |I| \times |O|$ matrix.

The last element in the tuple consists of *values*, $V(C_i \wedge I_j \wedge O_k)$, where the indices usually range over the consequents, interpretants, and objects in the ontology such that the values may be represented as a $|C| \times |I| \times |O|$ matrix.

We will say that such an interpretive ground is relevant to a relatively rational agent in a particular context insofar as the agent uses the ground to evaluate its instigations given its sensations in that context. And we will assume that the mode of evaluation may itself be modeled using some variant of the following equation:

$$V(I_i/S_j) = \sum_k \sum_l V(C_k \wedge I_i \wedge O_l) \cdot P(C_k/I_i \wedge O_l) \cdot P(O_l/S_j), \qquad (1.1)$$

where the last term in the equation may be expanded using Bayes' theorem:

$$P(O_l/S_j) = \frac{P(S_j/O_l) \cdot P(O_l)}{\sum_m P(S_j/O_m) \cdot P(O_m)} = \frac{P(S_j/O_l) \cdot P(O_l)}{P(S_j)}. \qquad (1.2)$$

Given a suitable specification of an agent's ground, equation (1.1) allows us to calculate the value of various interpretants (I_i), given various signs (S_j), as a function of the objects (O_l) that are stood for by those signs, as well as the consequents (C_k) that are created by those interpretants. As may be seen, its terms are fully specified by the elements in the interpretive ground, and hence by the agent's ontology, commitments, and values. Assuming that the agent instigates the most valuable interpretant that it can, given the sign that it presently senses, this equation thereby specifies the behavior of the agent in all relevant situations: how it will most likely interpret each of the possible signs it might encounter insofar as it is beholden to such a ground. The agent's sign-dependent interpretants—and thus their actions, inferences, interrelations, and affect—are thereby guided by such a ground.

Given its importance to the analysis, equation (1.1) will be referred to as the "main equation" in the chapters that follow. As may be seen, it involves a double summation. First, there is a sum over all the possible objects (as those entities that are stood for by signs); second, there is a sum over all the possible consequents (as those entities that are created by interpretants). In so doing, the agent is, in effect, summing over the product of values, eventualities, likelihoods, and priors for the objects and consequents in question. This double sum (in the discrete case), or double integration (in the continuous case to be discussed later), is our way of handling the two kinds of value that are essential to meaning: what signs stand for and what agents strive for.

It should be noted that we did not derive the main equation from first principles, or anything else so sophisticated. Rather, using classic notions like conditional probabilities, expected values, and Bayes' theorem, we simply expressed the foregoing conceptional framework (consisting of signs, interpretants, objects, consequents, agents, and grounds) in the simplest mathematical formulation we could imagine. Peirce himself would have had access to, and thus could have employed, the same resources for his own writings. There are, to be sure, many other ways that we might mathematically realize these relations, using more recent developments like neural networks, expected utility, and prospect theory, inter alia. We will look at some of these alternative formulations, which are arguably more complicated, and possibly more realistic, in later chapters. Phrased another way, this mathematical model is designed to be as simple as possible while remaining useful, plausible, transposable, and generalizable.

While the main equation can be used for any ontology of any size, in the particularly simple scenario that will be analyzed first, there will be two signs (S_1 and S_2), two objects (O_1 and O_2), two interpretants (I_1 and I_2), and two consequents (C_1 and C_2). In effect, then, we will be dealing with a universe in which there are 16 'possible worlds'—understood as the $2 \times 2 \times 2 \times 2$ possible ways that such a scenario-specific universe might be realized. We will later introduce a continuous form of this equation, in which there might be an infinity of possible signs, objects, interpretants, and consequents (and hence, loosely speaking, an infinity of possible worlds).

As will be discussed at length in later chapters, and as presupposed by our specification of the agent's commitments—$P(O)$, $P(S/O)$, and $P(C/I \wedge O)$—we typically assume that (the agent tacitly assumes that) objects causally influence signs, interpretants and objects causally influence consequents, and signs only causally influence interpretants through agents. Signs typically provide evidence of objects; they are auspicious. Interpretants, in the context of objects, typically create consequents; they are efficacious. This is what is meant when we say that an agent's ground takes for granted a certain causal structure (perhaps erroneously).

Just as grounds may be framed relatively subjectively (as an agent's understanding of the causal patterns and valuable resources present in an environment, insofar as such patterns and resources are relevant to the agent's semiotic processes), they may also be framed relatively objectively (as the casual patterns and valuable resources that are actually found in an environment, insofar as such patterns and resources influenced the semiotic processes of such agents and/or gave rise to them as agents). In other words, for certain kinds of agents, in certain contexts, we make sense of their semiotic processes by reference to their *subjective grounds* (such as their beliefs, desires, habits, or instincts); for other kinds of agents, in other contexts, we make sense of their semiotic processes by reference to the *objective grounds* of their environments (such as the conditions under which they evolved, or past regimes in which their behavior was socially regimented, governed, or otherwise punished and rewarded). As will be seen, to make sense of the behavior of most kinds of agents, both kinds of grounds must be considered, as well as the discrepancies and overlaps between them.

All this is another way of saying that analysis must take into account such interpretive grounds, be they relatively objective or subjective, in order to make sense of semiotic processes and the origins of the agents involved in them. Indeed, it is usually much more complicated than this, for the consequential environments of most semiotic agents are precisely the semiotic processes, and hence semiotic grounds, of other agents, which are themselves often temporally dependent and hence in transition. Thus there is often nothing like an objective context, or fixed environment, for agents to accommodate to—at least on longer time scales. As will be seen, the analysis is precisely designed to handle such complicated agentive couplings, and the complex time-dependent interactions that result.

It should be emphasized that the main equation is only interesting insofar as the scenarios that it is used to model are interesting. Indeed, it would be a simple thing to specify the ground and then solve for the best interpretant, given each sign. Except in the case of the first scenario analyzed, we will not engage in such calculations. We will rather examine solutions to this equation for various scenarios as particular variables change to highlight various critical points, or thresholds.

It should also be emphasized that the main equation, as it stands, is meant to be flexible, dynamic, and portable. Many scenarios will turn on relatively reduced variants of it (e.g., contexts in which consequents can be ignored, such that the main equation only makes reference to signs, objects, and interpretants). Most of the scenarios examined will involve more than one agent, such as a signer interacting with an interpreter. In such cases, the equations governing each agent's behavior may make reference to the equations governing the behavior of the other agents. Finally, insofar as the grounds of agents often transform over time, the equations will often be functions of time. In short, the main equation will often be reduced to simplified variants, coupled with complementary equivalents, and/or dynamically iterated.

Moreover, such equations will usually be put into relation with other mathematical approaches. In so doing, it will be seen that the values in question can range from economic values (like price and utility) and existential values (like prudence or honor) to biological values (like fitness), affective values (like pleasure and pain), and thermodynamic values (like free energy). We will thereby put this equation in dialogue with various insights and formulations from mathematical biology, statistical mechanics, information theory, political economy, machine learning, and anthropology. Indeed, as the title of this book suggests, and as later chapters will make clear, this model of meaning is really composed of a family of closely related models.

The next section highlights two of these alternative approaches, the first involving agents that evolve, and the second involving agents that learn. This will be our first foray into the dynamic coupling of semiotic grounds and semiotic processes, whereby each enables and constrains the other, leading to agents that alter their behavior, and thus their interactions with each other, through experience over time.

1.3 Agents That Evolve and Learn

While the main equation, as just described, is meant to model the behavior of relatively mindful agents, such as humans and animals, that *select* particular interpretant-sign relations given the contents of their grounds, a variant of it will also be used to model the effects of relatively mindless agents, such as alleles, which *are selected* as a function of the degree to which the semiotic processes that they engender are relatively value-generating, or fitness-promoting, in a given environment.

More precisely, we will examine alleles A_i that cause their bearers to cause (→) particular events (qua interpretants) in the context of particular events (qua signs) with certain conditional probabilities:

$$A_i \rightarrow P_{A_i}(I/S)\,, \tag{1.3}$$

where the relative fitness of such alleles may be determined using the following equation:

$$F(A_i) = \sum_{jk} P(S_j) \cdot P_{A_i}(I_k/S_j) \cdot V(I_k/S_j)\,. \tag{1.4}$$

Here, the third term on the right-hand side is simply the main equation, whose components will be suitably reframed (as discussed previously) to describe relatively objective patterns and resources within an environment (as opposed to agent-specific, and hence often relatively erroneous and/or subjective, understandings of such patterns and resources). And the first term on the right-hand side, understood as the probability of a triggering event, qua sign, may depend on the features of a relatively stable environment or the behaviors of other evolving agents, inter alia.

This measure of fitness will be used to track the changing frequency of such alleles over generations within a population, such that the evolution of biosemiotic agents may be studied. In particular, it is argued that the genomes of organisms embody interpretive grounds that transform on phylogenetic time scales, whereby the behavioral phenotypes generated by such grounds become better adapted to particular environments (however fleetingly), and where such environments can include the semiotic processes and interpretive grounds of other agents.

The main equation can also be used to study semiotic agents that learn through experience via positive and negative regimentation, reinforcement, or discipline. Such agents register the results of past experience in their habits, as opposed to their genomes. To see how, let the term *practice* denote the instigation of an interpretant in the context of a sign. And let the term *register* (e.g., a 'habitus' in the tradition of Marcel Mauss, or simply a 'habit' in the tradition of Peirce) denote an array of weights indexed to particular practices $W(I/S)$. The weights in an agent's register, or habitus, are not just determined by its past experience; they are also determining of its future practices. In particular, each time an agent engages in a practice (instigating some interpretant in the context of some sign), the value that it receives is added to the appropriate weight in its register; and the greater the weight of a practice in the agent's register, the more likely the agent is to instigate that particular interpretant in the context of that particular sign in future interactions.

As will be more carefully shown later, if each of the weights in an agent's register at some slice of time is known, $W(I_i/S_j)$, the expected weight of each of the behaviors in their register during the next slice of time, $W'(I_i/S_j)$, can be determined using the following difference equation:

$$W'(I_i/S_j) = W(I_i/S_j) \cdot e^{-\alpha} + P(S_j) \cdot P(I_i/S_j) \cdot V(I_i/S_j)\,. \qquad (1.5)$$

Here, α is a parameter that determines how rapidly the agent 'forgets' prior experience (by discounting weights that were registered earlier); $P(S_j)$ is the probability that a particular sign occurs in some environment, which is analogous to the same term in equation (1.4); and $V(I_i/S_j)$ is again the main equation. The term $P(I_i/S_i)$, meaning the probability that the agent instigates I_i in the context of S_j, may be determined by the current weights in the agent's register:

$$P(I_i/S_j) = \text{softmax}\big(W(I_i/S_j)\big) = \frac{e^{\beta \cdot W(I_i/S_j)}}{\sum_k e^{\beta \cdot W(I_k/S_j)}}\,. \qquad (1.6)$$

Here β is a parameter that determines how concentrated the probabilities are around the highest weights in the agent's register, and the softmax function takes in weights and turns out probabilities (by exponentiating all the weights in the register and then normalizing the weights in each column). In short, such agents internalize the effects of prior interactions, and the habits thereby engendered guide their future practices.

We now turn to coupled semiotic agents that come to communicate using conventional signals.

1.4 Signers, Interpreters, and Signals

Both of the foregoing kinds of agents—those that evolve and those that learn—often arise in contexts that involve multiple agents, each of which plays a distinct and complementary role in a division of semiotic labor. Figure 1.3 shows one such scenario.

As may be seen, there are two agents mediated by a single semiotic process, consisting of an object (O), a sign (S), and an interpretant (I). The signer, A_1, senses an object and instigates a sign. The interpreter, A_2, senses the sign and instigates an interpretant. And the relation between the object and interpretant yields a value that benefits both signer and interpreter alike.

Just as the interpreter's access to the object is mediated by the signer, so is the signer's access to the interpretant mediated by the interpreter. Each agent, as it were, has direct access to only two-thirds of the interaction; so

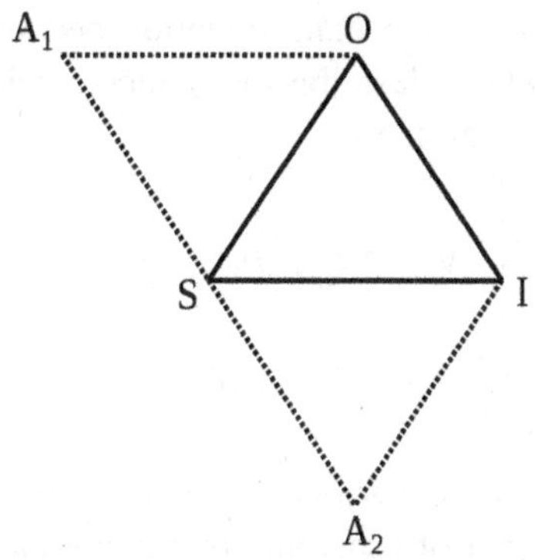

Figure 1.3: Signaling Agents

the behavior of each agent makes sense only in reference to the behavior of the other.

While such a scenario may be imagined in many different ways, for present purposes the object should be understood as the location of a source of food that the signer can sense but not obtain (and the interpreter can obtain but not sense); the sign as some sound or gesture that the signer can instigate and the interpreter can sense; and the interpretant as the action of going to one location or another to obtain the food (which will then be shared by the two agents).

Note that in such a simple scenario, there is no need of consequents. In effect, it is assumed that what the agent instigates (going to some location) is what the agent achieves (arriving at that location).

To flesh out the interpretive grounds of the two agents and make the scenario relatively general (and symmetric), we might assume that there are N objects, N signs, and N interpretants. We might assume that each of the N objects is equally likely, such that $P(O) = 1/N$. We might assume that the semiotic strategies of the signer and interpreter are $N \times N$ matrices, understood as time-dependent, conditional probabilities, which may be denoted as $P_t(S/O)$ and $P_t(I/S)$, respectively. And we might assume that the value received by each agent, as a function of their signifying and interpreting practices, is

$$V(I_i \wedge O_j) = \frac{v}{2} \cdot \delta_{ij}, \tag{1.7}$$

where v is the value of the food and δ_{ij} is the Kronecker delta function (or an $N \times N$ identity matrix), such that δ_{ij} is 1 when $i = j$ and 0 otherwise. In other words, so long as the interpreting agent goes to the right location, regardless of the sign that is employed by the signifying agent, both agents benefit equally.

Given such assumptions, the main equation becomes two coupled, time-dependent equations, which describe the values received by the agents as a function of their semiotic strategies:

$$V_t(S_i/O_j) = \sum_k V(I_k \wedge O_j) \cdot P(O_j) \cdot P_t(S_i/O_j) \cdot P_t(I_k/S_i)\,, \tag{1.8a}$$

$$V_t(I_k/S_i) = \sum_j V(I_k \wedge O_j) \cdot P(O_j) \cdot P_t(S_i/O_j) \cdot P_t(I_k/S_i)\,. \tag{1.8b}$$

As may be seen, for the scenario in question, we assume that the values and priors are independent of time, and that consequents can be ignored. As may also be seen, the value of each agent's strategy depends on the other agent's strategy. Indeed, given the way that $V(I \wedge O)$ was defined, the expressions being summed over in $V_t(S/O)$ and $V_t(I/S)$ are equivalent. This should be unsurprising: the fates of such semiotically coupled agents rise and fall together.

To be sure, this is one of the simplest kinds of semiotic systems: two agents learning, or evolving, to employ relatively conventional (or 'arbitrary') signals. If plugged into the kinds of selection processes described in section 1.3 (which themselves are expanded to encompass signers in addition to interpreters), the two agents will evolve, or learn, to employ one of the $N!$ possible codes to communicate. It is introduced here to show how the main equation, or rather one of its simplified variants, can capture one of the most celebrated of semiotic processes.

We have so far offered a relatively capacious definition of meaningful interrelations, shown how such interrelations can be studied mathematically (by projecting a certain causal, statistical, and evaluative structure onto them), highlighted some of the presuppositions that go into such a projection, contrasted three kinds of prototypic agents (those that reason, those that evolve, and those that learn), and sketched some of the ways that the main equation can be used to capture the division of semiotic labor and processes like signaling.

The next section examines the relation between three tightly coupled, and easily conflated, concepts.

1.5 Meaning, Information, and Value

Terms like *meaning*, *information*, and *value* are somewhat slippery, sometimes referring to what signs stand for, and sometimes referring to what agents strive for. We now sketch some of the ways that these terms will be used, and mathematically realized, in the chapters that follow. To keep things simple, we will focus on relatively humanlike agents, and let later chapters generalize such claims to a wider range of agents. But be warned: as was the case in the last two sections, a few parts of this discussion may seem relatively elliptic at first glance, insofar as the concepts and measures in question will not be fully explained until certain ideas and definitions are in place.

In a very narrow sense, the meaning of a sign is the object that it stands for. To paraphrase Peirce, the object of a sign is what a sign, so far as it is known, and known as a sign, allows one to know. But, as seen in the foregoing sections, this term can also be expanded to include the motivation of the signer (for expressing the sign in the first place), the motivation of the interpreter (for offering the particular interpretant that they do), the interpretant per se, the consequent that the interpretant effects, relations between relations more generally (recall figure 1.2), and much else besides. Focusing on the narrow sense of this term, we will see that signs rarely have a single well-defined meaning. What they *do*, rather, is change an agent's commitments regarding the probabilities of objects (and thereby change the likelihood of particular interpretants). Moreover, objects themselves are often best understood not as particular entities or events, but as sets of possible worlds.

The foregoing points might be formalized by thinking of the meaning of any sign, S_i, as a row of vectors (one for each object, O_j, in the agent's ontology), where the tail of a vector is located at $P(O_j)$, understood as the prior probability of O_j, and the head of the vector is located at $P(O_j/S_i)$, understood as the posterior probability of O_j (where the transition from prior to posterior probability is conditioned on the occurrence of the sign; see figure 1.4). In effect, such sign-specific sets of vectors, which are thus tensor-like, represent the change in the probabilities of objects given signs (at certain times). A sign not only may make a particular object more or less probable (to an agent) than before, it also may do so to a greater or lesser degree. In this way, the meaning of a sign is not only agent specific and time dependent (insofar as it turns on the present interpretive ground of a particular agent), but also graduated and modal (insofar as it turns on relative intensities of various propensities).

Crucially, and keeping with caveats introduced in earlier sections, the foregoing points are not meant to imply that any agent is necessarily, or even usually, conscious of such changes. It is simply the case that it is necessary to take into account such changes to make sense of semiotic processes insofar as the grounds of agents and environments depend on them.

For the purposes of this discussion, a possible world is any conjunction of an object, sign, interpretant, and consequent that is available in an agent's ontology. Let w_{ijkl} denote the world in which the object is O_i, the sign is S_j, the interpretant is I_k, and the consequent is C_l. If a universe, denoted as Ω, is a complete set of possible worlds, then the size of this set, its cardinality, is simply

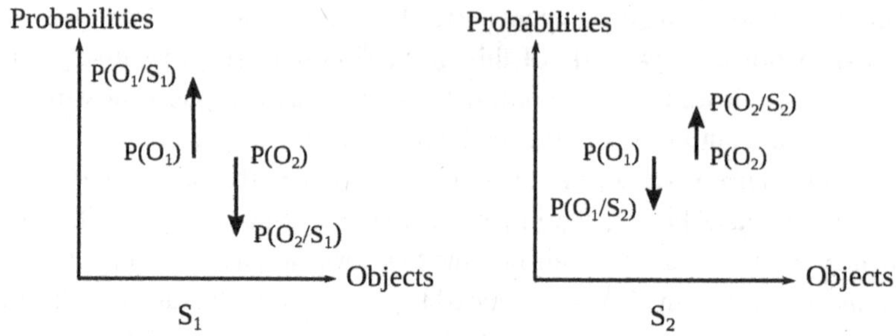

Figure 1.4: Meaning of Signs as Change in the Probabilities of Objects

$|O| \times |S| \times |I| \times |C|$, where $|O|$ is the number of objects in the agent's ontology (and similarly for the other three variables; see figure 1.5). The agent's ontology is really a cosmology—not so much a view of the world (or *Weltanschauung*) as a stance towards a universe (in the context of a scenario). And a possible world is simply one way in which such a universe might unfold, and thereby become fully realized.

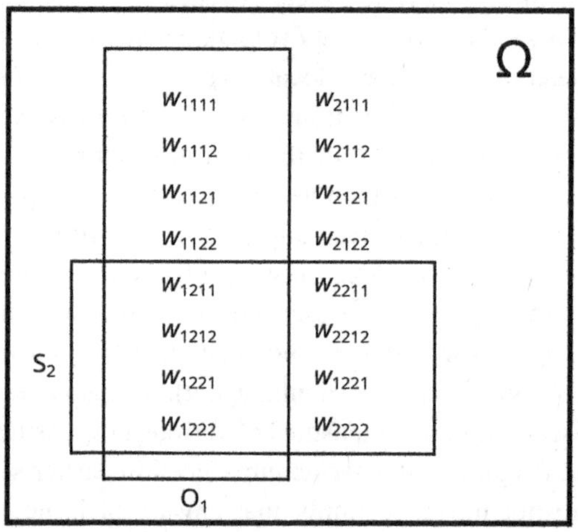

Figure 1.5: Object and Sign as Sets of Possible Worlds

In saying that an object is a set of possible worlds, we mean that it is the set of worlds compatible with the index of the object in question. It is thus all the worlds in which the object obtains (and hence certain ontology-specific, and hence ground-specific, conditions are met), however else those worlds may otherwise differ. For example, O_1 is the set of worlds whose first index is fixed at 1 (and whose other indices are free to vary). Crucially, if an object is a set of worlds, so are signs, interpretants, and consequents. For example, S_2 is the set of worlds in which the second index of w_{ijkl} is fixed at 2 (while all the other indices are free to vary). To return to the first point, learning that one is in a world in which S_2 obtains changes one's estimation of the probability that one is in a world in which O_1 obtains. And this, in turn, changes how one acts (thereby changing the probability that a particular interpretant obtains) which, in conjunction with the object, changes the probability that a particular consequent obtains.

What signs do, in this framing, is change the probability that the actual world in which an agent resides belongs to one subset of the universe or another. Phrased in another way, insofar as a sign is meaningful for an agent, in the narrow sense, by knowing something about the second index of w_{ijkl}, that agent can come to know something about the first index (which thereby constrains the other two indices as well). Loosely, meaningfulness allows an agent to skip across slashes, now framed as gaps between available indices on possible worlds.

As will be seen, one kind of interpretant enabled by the meaning of a sign—perhaps the *ultimate* interpretant—is precisely a transformation of the agent's ground, which may include a change in their ontology, and hence their beliefs and values, as well as their habits more generally. It is, in effect, a new stance towards a universe (be it embodied or enminded), and all the worlds that it may include or omit, reject or embrace. We borrow Peirce's terminology, however different our theory is, because such a change in habit or belief, as engendered by a sign, may potentially affect all future semiotic processes in which the agent participates, and hence all future worlds that the agent may one day inhabit.

Within this framing, a somewhat subtle point is that signs and objects, as well as interpretants and consequents, are not usually entities or events per se (such as a symptom or a sickness); they are, rather, interrelated entities and events such as 'the person evinces the symptom' (as a sign) or 'the person is sick' (as an object). In effect, such variables are full-fledged propositions. In one sense, then, they are sets of worlds; in another, equivalent sense, they are functions from worlds (and indices more generally) to truth values. That is, they return the value 'true' (as output), if the world that they are given (as input) is a member of the set that they constitute. Moreover, as propositions, they are infinitely rich in how they might be rendered—and so they may involve complicated relations not just among entities, events, qualities, and the like, but also among times and worlds per se. Just as anything sensible can be a sign and anything instigatable can be an interpretant, just about anything imaginable can be an object or a consequent. At least *in potentia*, to a particular kind of agent, with a large-enough universe (given its ontology, and semiotic capacities more generally).

We now move from the meaning of a sign (in a narrow sense) to the information of a sign, which is to be understood as the change that the sign makes in an agent's uncertainty regarding which object—or which world more generally—obtains. In one important sense, which goes back to Claude Shannon, the informativeness of a sign, S_i, is its surprise value, understood as the logarithm of its improbability: $-\log_2 P(S_i)$. The informativeness, or 'entropy' (denoted as H), of a set of signs is then the average surprise value of all the signs in the set: $H(S) = -\sum_j P(S_j) \cdot \log P(S_j)$. Rather than focus on the surprise value of a set of signs, however, we are interested in the expected *change* in the surprise value of a set of objects due to the acquisition of a sign (for an agent, given its ground, on average). Somewhat loosely, this may be understood as the average number of questions that a sign saves an agent from having to ask (to learn the identity of the object), and hence the change in the agent's uncertainty regarding which object (or set of worlds) obtains:

$$\mathcal{I}(S) = \sum_j P(S_j) \cdot \left[H(O) - H(O/S_j)\right]. \tag{1.9}$$

Here,

$$\begin{aligned} H(O) - H(O/S_j) = &-\sum_i P(O_i) \cdot \log_2 P(O_i) \\ &+ \sum_i P(O_i/S_j) \cdot \log_2 P(O_i/S_j) \end{aligned} \tag{1.10}$$

is the difference between the prior and posterior uncertainties of the agent regarding which object obtains. (The posteriors, $P(O/S)$, may be found from the priors and likelihoods using Bayes' theorem.) In effect, one simply calculates the average surprise value of all the objects in the set (prior to the acquisition of a sign) and subtracts from this the average surprise value of all the objects in the set conditioned on the occurrence of a sign, which itself is averaged over all possible signs that could occur. It can be readily shown that this is equivalent to the relative entropy of $P(O/S_j)$ with respect to $P(O)$, as averaged over all possible signs.

A related measure may be used to calculate how the presence of an agent contributes to the change in Shannon entropy (or 'complexity') of a universe, in the sense that their presence makes certain modes of patterning more or less likely. This is because information is closely related to entropy, just as organization, and hence the reduction of entropy—understood as the creation of patterning—is closely related to work. This is not work in the sense of giving form to substance for the sake of function, but work in the sense of organizing complexity for the sake of predictability. Indeed, the possible worlds of formal semantics are closely related to the microstates of statistical mechanics, and this fact allows for a relatively seamless connection not only between meaning and information, but also between Boltzmann entropy, organization, work, and constraint. Phrased in another way, besides thinking about the meaning, information, and value of a sign (from the standpoint of an interpreting agent), one may also consider the entropy (complexity or unpredictability) of a scenario—itself filled with one or more interpreting agents—from the standpoint of an analyst.

Having distinguished between the meaning and information of a sign, in a relatively specific sense, we may now take up the value of a sign, which may be understood as how much an agent would be willing to sacrifice to obtain the sign, insofar as its meaning allows that agent to better guide its actions (justify its reasons, express its feelings, and/or modify its social relations).

If the interpretive ground of an agent is known, the main equation allows one to calculate the value of an interpretant given a sign, $V(I_i/S_j)$. As was shown earlier, this equation takes into account not only the object that the sign stands for, but also the consequent that the interpretant leads to. Indeed, using the agent's ground, one can even calculate $V(I_i)$, which may be understood as the value of an interpretant in the absence of any information provided by a sign. (Simply replace the posteriors, $P(O/S)$, in equation (1.1) with the priors, $P(O)$.) Assuming that the agent always instigates the most valuable interpretant that it can (given its ground), whether it has received a sign or not, one can calculate the change in expected value that any particular sign provides, $\mathcal{V}(S_j)$. Averaging this value across all possible signs, one can calculate the expected value of receiving a sign per se:

$$\begin{aligned}\mathcal{V}(S) &= \sum_i P(S_i) \cdot \mathcal{V}(S_i) \\ &= \sum_i P(S_i) \cdot \left[\max\big(V(I/S_i)\big) - \max\big(V(I)\big)\right].\end{aligned} \tag{1.11}$$

Here, the term $\max\big(V(I)\big)$ denotes the expected value of the most valuable interpretant (in the absence of any sign), whereas the term $\max\big(V(I/S_i)\big)$ denotes the expected value of the most valuable interpretant given some particular sign. As may be seen, $\mathcal{V}(S_i)$ is just the difference in expected value that sign S_i makes (insofar as it is meaningful to an agent). And $\mathcal{V}(S)$ is just the average difference in expected value that signs provide an agent, insofar as they allow that agent to better relate to its environment, and thereby better direct its efforts.

Note that, in contrast to the information of a sign, $\mathcal{I}(S)$, which only takes into account the relation between objects and signs (in the guise of priors and likelihoods), the value of a sign, $\mathcal{V}(S)$, takes into account all aspects of an agent's ground (and hence all relations among objects, signs, interpretants, and consequents). Indeed, a key argument of chapter 2 will be that standard measures of information—like relative entropy or Kullback–Leibler divergence—are relatively uninformative.

As will be seen, one can even calculate the value of an agent's interpretive ground, $\mathcal{V}(G)$, given the actual causal patterns and valuable resources available in an environment. Somewhat loosely, this is a measure of the goodness-of-fit between the agent's stance towards a universe and the universe per se. Such a measure allows one to assign a relative value to better and worse representations of reality, and hence to weigh the costs and benefits—whatever the currency, be it euros, pleasure, or free energy—of having a particular worldview.

If the use-value of a sign may be understood as the function that it serves in a communicative encounter, then three additional and closely related varieties of value have just been provided: truth value (qua meaning), surprise value (qua information), and exchange value (qua price). As should be clear, while such terms have very different measures, they all turn on relatively quantitative changes that are induced by the presence of signs in relation to agents and their interpretive grounds in certain worlds, at certain times. Each of these concepts will be expanded and explicated, as well as qualified and critiqued, in the chapters that follow.

The next section surveys the arguments and architecture of this book.

1.6 Overview of the Chapters

In broad strokes, all the chapters that follow make two kinds of contributions. At one level, they offer case studies of relatively simple, adjustable, and generalizable scenarios. In so doing, they provide a concrete exploration of the entailments of this model, walk readers through the mathematical details, and provide lively demonstrations of interesting systems that may be easily adapted to wider concerns. At another level, they flesh out the presumptions of the model or show how it can be extended to understand more complicated processes. In so doing, they not only explicate, but also criticize, complicate, and reformulate the conceptual underpinnings of the model and its mathematical realization.

Aside from this introduction and a brief conclusion, this book is composed of four parts, each of which consists of two chapters. The chapters in Part I analyze the behavior of relatively mindful agents that can flexibly determine their own behavior (within certain limits).

Chapter 2 analyzes the relation between sickness, symptoms, treatments, and outcomes. It models the behavior of agents that need to decide whether they should take a medication, given the results of a diagnostic and the possibility of side effects. It also calculates the value of a sign by determining how much such agents would be willing to pay for the diagnostic, given their grounds. It argues that standard measures of information, such as Shannon entropy (which only takes into account the distribution of signs) and Kullback–Leibler divergence (which only takes into account the distribution of signs and objects), are wholly inadequate for understanding meaning. In their stead, it offers a way to measure the value of meaningful interrelations: one that takes into account signs, objects, interpretants, and consequents; and not just the ways that such components are causally interrelated, but also the ways that such interrelations are valued by an agent. And it argues that a good interpretive ground—one that adequately represents the patterns and values in an environment—is one of the most valuable goods. This chapter thereby retheorizes seemingly microeconomic processes in meaningful terms.

Chapter 3 studies the real-time dynamics of predators chasing prey in an open environment. In contrast to chapter 2, which turns on the discrete version of the main equation, this chapter uses the continuous version of the main equation: in effect, the coupled movements of predators and prey are simultaneously signs for each other to interpret, and interpretants of each other's signs. It analyzes the interactions of such agents in spaces of various dimensions and environments with various terrains. In so doing, it shows how this framework can be used to study real-time, coupled, and continuous interactions between two or more semiotic agents, as they take turns inhabiting the roles of signer and interpreter. It thereby studies the dynamics of relatively simple conversations between agents who are trying to either capture or elude each other. The main equation, in its continuous form, thereby functions as a kind of kernel that generates complex world lines, understood as the coupled movements of signers and interpreters through space-time.

Unlike the chapters in Part I, which focus on relatively mindful agents, and hence agents who can choose their own actions, as well as reason more generally, the chapters in Part II focus on agents that evolve on evolutionary time scales.

Chapter 4 introduces readers to the mathematical machinery of difference equations and critical points, as a key means for tracking the evolutionary dynamics of such biosemiotic agents. After working through several simple scenarios, whereby agents evolve in relatively fixed environments, or evolve in dynamic environments constituted by other evolving agents, it works through the details of an extended case study. In particular, it analyzes a scenario in which two populations of agents inhabiting a relatively noisy, but value-rich, environment come to share a code. It measures the value of particular codes insofar as such codes better enable their users to communicate about salient features of their environments to cooperatively secure the values that are afforded by those environments. And it shows how different kinds of codes are more or less likely to emerge in different kinds of environments.

Chapter 5 returns to a classic scenario from evolutionary game theory: pairwise competitions over a valuable resource between relatively aggressive hawks and relatively passive doves. It shows what would happen to an otherwise stable population of hawks and doves if a particular kind of mutant is introduced: an agent that can semiotically ascertain—if only slightly better than chance—the identity of its competitor (hawk versus dove) and alter its behavior accordingly. It studies the conditions in which such a mutant is sieved out of the population completely, drives out one or both of its competitors, or comes to stably exist with a certain frequency in the population. In effect, this chapter studies the evolutionarily stable strategies of such social parasites as a function of their semiotic capacities.

Unlike the chapters in Part II, which focus on the evolution of semiotic agents, the chapters in Part III focus on semiotic agents that can learn on interactional time scales.

Chapter 6 models the behavior of agents that can improve the fit between their interpretive grounds and the environments they inhabit by internalizing the consequences of their past practices and adjusting their present practices accordingly. It reframes reinforcement learning and Edward Thorndike's law of effect in terms of the main equation. And it uses this framing to analyze several key scenarios: when agents must learn a set of purely arbitrary conventions to communicate with each other; when the conventions learned are relatively motivated, such that the code privileges certain signs over others; and when a parasite (third party or interloper) benefits from such conventions, such that the agents in question must repeatedly adjust their code to thwart it. It thereby explores the relation between nature and convention (or the motivated and the arbitrary) and the relation between codes and ciphers (or messages and secrets).

Chapter 7 models semiotic agents as neural networks that are able to learn from experience. It shows that the architectures of such networks, along with the values of the parameters they contain, are equivalent to interpretive grounds. It thereby puts this model of meaning in dialogue with classic ideas from machine learning. After introducing readers to artificial neural networks and machine learning techniques, it uses those techniques to model a scenario in which relatively simple agents establish a shared code consisting of N arbitrary conventions. It argues that backpropagation—the key algorithm used to establish parameter values in a neural network—is a mode of meta-semiosis. And it shows the relation between language models (in a narrow sense) and models of meaning that are theorized in relation to models of reality. Loosely, if the former focus on word-word relations (in the guise of next-word prediction), the latter focus on word-world relations (in the guise of sign-object and interpretant-consequent mediation). It argues that the key capacity of language models (next-word prediction) is their main limitation: *worldlessness*.

Unlike the chapters in Parts I, II, and III of this book, which focus on case studies of revealing scenarios, whether they involve selecting agents or selected agents, the chapters in Part IV explore some of the deeper presuppositions and farther-reaching entailments of the model.

Chapter 8 introduces readers to the technical details of possible worlds, an analytic approach that is often used to model the semantics of natural languages. It introduces the idea of a possible world using standard conventions, and then it connects possible worlds to formal semantics, with a particular focus on modal operators (such as necessity and possibility). It relates possible worlds to this mathematical model of meaning, with a particular focus on Bayesian networks (or causal graphs) and expected value. In so doing, this chapter shows precisely what kind of causal graph was presupposed by our model, and it derives the main equation from first principles. Finally, this chapter shows the relation between times and worlds and uses this relation to understand the changing grounds of agents as they come to know—and transform—the worlds around them. In some sense, then, it offers an account of possible world semiotics, which is meant to encompass not only possible world semantics, but also the pragmatics of actual worlds.

Chapter 9 uses the main equation to model meta-semiotic processes, whereby the grounds of semiotic agents are figured, and thereby become the objects and consequents, and hence topics and ends, of semiotic practices. It shows how agents may update their interpretive grounds by changing not just their commitments and values, but also their ontologies and models, as a function of the feedback that they receive regarding the fit between their interpretive ground and the environment they inhabit. It also shows the ways that agents may come to model the interpretive grounds of other agents, as well as have their own interpretive grounds modeled in turn, to higher and higher levels of embedding. It argues that culture may be usefully understood as an intersubjectively held interpretive ground, along with the semiotic processes, qua figured practices, that constitute its roots and fruits. And it further explores the range of relatively antagonistic agents—such as enemies, parasites, and noise—that exploit and degrade the otherwise shared interpretive grounds of others.

Finally, a brief conclusion lays out the relation between interpretive grounds and Max Weber's notion of an ideal type. It analyzes the relation between the interpretive grounds of analysts (who seek to model particular scenarios) and the interpretive grounds of the agents in the scenarios so modeled. Also, it highlights the productive tension between revelation and confrontation.

1.7 Relevant Literature

As discussed earlier in this chapter, this formulation of semiotic processes takes inspiration from Peirce (1984 [1867], 1986 [1877], 1986 [1878]). We borrow his notions of sign (or representatum), object, and interpretant; and add to them the notions of agent and consequent. And rather than treat interpretive grounds in terms of iconic, indexical, and symbolic relations, they are framed in terms of commitments, values, and ontologies. In this sense, we build on Peirce's notion of a guiding principle, while framing it in a way that explicitly takes quantities, or differential intensities, into account. Finally, as will become clear in the rest of this section, while Peirce had a deep understanding of core topics in probability and statistics, we make extensive use of mathematical ideas that either preceded him (like Bayes' theorem and expected value) or came after him (such as Bayesian networks and game theory).

The body of secondary literature on Peirce is enormous. Three particularly helpful guides to his arguments and interpretations of his ideas may be found in Colapietro (1988), Parmentier (1994), and Lee (1997). These authors also develop important connections between Peirce's ideas and classic works in linguistics, anthropology, and philosophy.

Kockelman (2013a) offers a more detailed presentation of the framework outlined in section 1.1. Kockelman (2017) extends this framework to think about Bayesian reasoning, Kockelman (2020a) uses it to analyze machine learning, and Kockelman (2024) uses it to understand large language models and artificial intelligence. In contrast to the analysis offered in this book, however, these works are more directly addressed to anthropologists, critical theorists, and scholars of science and technology.

For more on the slash, as framed through metaphors like bridges, horizons, and fog, see Kockelman (2010, 2016, 2024). For classic work on similar themes, see Whitehead (1920) on jagged edges (coming out of Peirce and Hume), Bateson (2000) on slash marks (coming out of Claude Shannon's notion of redundancy), Roman Jakobson on the backslash, E^N/E^S, which relates the narrated event and the speech event (coming out of Otto Jesperson's notion of shifters), Thomas Bayes on the probability of a hypothesis conditioned on evidence, $P(H/E)$, and of course folks like Albert Einstein and Stephen Hawking on the contrast between the horizons of subjects and the world lines of objects.

The mathematical model offered here, as sketched in section 1.2, incorporates two relatively simple, yet powerful, mathematical ideas: Bayes' theorem (Bayes 1763; Laplace 1951; Jeffreys 1998) and expected value (Laplace 1951; Huygens 1714). Indeed, both ideas were already well developed by Pierre-Simon Laplace, whose celebrated text *Théorie analytique des probabilités* was published in 1814 (and thus about 25 years before Peirce was born). Hacking (2001, 2006) offers a particularly clear and inspired introduction to the philosophy of probability, the rise of statistics, and Peirce's relation to it.

While the main equation was introduced using expected value, it will be extended to encompass expected utility. And while our framing of certainty is left unspecified, we often presume subjective probabilities (as least when dealing with subjective grounds), following scholars like Ramsey (2016), de Finetti (2017), and Savage (1972). Savage's classic work, *The Foundations of Statistics*, which puts together Bayesian inference and expected utility and even offers a prescient account of the utility of information, is particularly relevant.

Other approaches to value, often highly critical of expected utility, include prospect theory (Kahneman and Tversky 2013; Levy 1992; Barberis 2013), Veblen-inspired theories of distinction (Veblen 2017; Bourdieu 1987, 1977), and ideas like strong evaluation (Taylor 1992), value rationality (Weber 2019), and higher-order desire (Frankfurt 1971). While our approach to value, at least in chapter 2, is relatively microeconomic, the framework is compatible with core ideas in substantivist economics (Malinowski 2013; Polanyi 2001; Sahlins 2013), as well as economic anthropology (Maurer 2011; Lee and LiPuma 2004; Lee 2020; Guyer 2004; Peebles 2010) and economic sociology (Lépinay 2011; MacKenzie, Muniesa, and Siu 2007). Kockelman (2007a, 2007b, 2015, 2020a) links the Peircean framework, as laid out in section 1.1, to many of these traditions.

This mathematical model is tightly coupled to diagrammatic representations, as intimated in the foregoing sections and as will be more fully fleshed out in chapter 9. For more on causal graphs, in relation to probabilistic reasoning, see the classic works of Sewall Wright (1921, 1934). Bayesian networks, as developed in chapter 8, are introduced and theorized in Pearl (1988) and Neapolitan (1990). On auspiciousness versus efficatiousness, see Weinrich (2020) and the literature that he cites (Lewis 1981; Skyrms 1982). For a survey and synthesis of the large literature on probabilistic causality, see Hitchcock (2021); for an early classic work, see Reichenbach (1956). A standard work on agents, artificial intelligence, probabilistic reasoning, and decision theory is Russell and Norvig (2002). For an engaging introduction to the history of ideas in these traditions, see Pearl and Mackenzie (2018).

This approach to information, entropy, and organization is grounded in key ideas from thermodynamics and statistical mechanics (Reif 1965), as well as classic ideas in information theory from works like Shannon (1948), Kullback and Leibler (1951), and Lindley (1956). We also employ the definition of organization in relation to constraints that was offered by Brooks and Wiley (1988). Kockelman (2013b) compares and contrasts classic theories of information through a pragmatist lens.

Our model of natural selection, via game theory, has its origins in Smith (1982). See Nowak (2006) for a more recent survey and synthesis of the field. For thinking about difference equations, and their fixed points, Strogatz (2018) is a key resource. And the model of organism-environment interactions offered here is resonant with Lewontin's classic text (1983) and allied work, such as the essays collected in Oyama (2003).

Reinforcement learning goes back to Thorndike (1898), as developed by scholars like Bush and Mosteller (1955), Roth and Erev (1995), and Herrnstein (1970). Skyrms (2010) helpfully summarizes high points in the history of reinforcement learning, and deploys those ideas in his account of signals. Our approach to machine learning is grounded in the classic paper by Rumelhart, Hinton, and Williams (1986). Nielsen (2015) offers an elegant and approachable introduction to this fascinating topic.

Our account of possible world semiotics builds on the possible worlds literature (Lewis 1986; Carnap 1988), especially as elaborated by formal semanticists (Kratzer and Heim 1998; Von Fintel and Heim 2011). The work of Angelika Kratzer (2012) is particularly relevant. Arguments offered in chapter 8 build on her notions of necessity, possibility, and conversational backgrounds. The concept of a possible world goes back at least as far as Gottfried Wilhelm Leibniz, as described in his *Theodicy*, and later in his *Monadology*. For more on his ideas, in relation to the world in which they germinated, see Borowski (2024) and Weatherby (2016).

Biosemiotics is a thriving discipline. The essays collected in Favareau (2010) nicely showcase the key texts and rich conceptual field of its proponents and originators, as does the book by Barbieri (2008) and the programmatic essay by Kull, Deacon, Emmeche, Hoffmeyer, and Stjernfelt (2009). That said, while the approach taken here is loosely aligned with these authors, insofar as it also takes Peirce, as well as von Uexküll (2013), as sources of inspiration, and insofar as it takes up the relation between semiotic processes and natural selection, the framing is mathematical rather than biological; and our overall conceptual approach is different as well. Moreover, we stick close to modern developments in mathematical biology, dynamic systems theory, ethology, linguistics, anthropology, cognitive science, and allied fields. Finally, as should be clear from this introduction, the model offered here applies to much more than biosemiotic processes.

For more on the difference between teleology, teleonomy, and teleomaticity, see Mayr (1974, 1992) and allied work (Enfield and Kockelman 2017). For classic accounts of animal signaling practices in relation to selecting processes, see Smith and Harper (2003) and the large discussion that it generated (Stegmann 2005). Millikan (2004) offers a particularly clear and influential discussion of meaning in relation to selection and the nature of function (or purpose) more generally.

Recent books by Deacon (2011) and Tomlinson (2023) are particularly resonant with this one, insofar as they approach meaning through a partially Peircean lens. They are, however, oriented toward the emergence of meaningful behavior in an extended historical sense (e.g., how mind might have emerged from matter, and/or what does and does not count as symbolic behavior as we look across all life forms). And they are qualitative accounts of such origins as opposed to mathematical models. Tomlinson offers a sharp critique of teleosemantic theories of meaning (such as those put forth by Millikan), teleodynamic theories of meaning (such as those put forth by Deacon), and relatively simplistic post-humanist theories of meaning (which currently abound in the social sciences and humanities). He also offers a very interesting distinction between meaning and information, and a detailed explication of the abstract machines that underlie the emergence of life. And both theorists utilize Peirce's trichotomy of icon, index, and symbol, which plays only a small role in this work. That said, this work is in agreement with Deacon, and especially thinkers like Kauffman (2003), Brooks and Wiley (1988), and Nelson (2004), that constraints are fundamental for thinking about entropy, information, and work. See, for example, Kockelman (2009). And it is allied with Tomlinson insofar as it is in dialogue with approaches in continental philosophy and critical theory, not just analytic philosophy and evolutionary biology. See, in particular, Kockelman (2011).

Other mathematical approaches to meaning, particularly resonant with this work, include Lewis (2008), Dretske (1981), and Skyrms (2010). Lewis offers a seminal account of the origins of conventions using game theory. Dretske synthesizes information theory and core concerns in analytic philosophy, focusing on flows of information in relation to knowledge. Skyrms builds on both these thinkers by working out a sophisticated account of signaling processes in relation to information content, reinforcement learning, game theory, and social networks. Unlike Lewis and Dretske, however, he also engages in extended mathematical modeling, summarizing work that was undertaken by himself, his collaborators (Hofbauer and Huttegger 2008; Huttegger 2007; Huttegger et al. 2014), and other theorists.

While this book builds on these three theorists, as well as the ideas of Smith and Millikan, it differs in several important ways. First, it is grounded in pragmatism, anthropology, and critical theory as much as analytic philosophy, game theory, and evolutionary biology. In contrast to Skyrms, it focuses on rational agents and evolving agents as much as on learning agents. It takes into account value as much as meaning, and entropy in addition to information. Unlike Skyrms, who summarizes work done elsewhere, this work walks readers through the mathematical model and its results in detail. Its earlier chapters are designed to get readers up to speed on key ideas and techniques from game theory, possible world semantics, mathematical biology, and information theory. And finally, the actual model that it offers, both conceptually and mathematically, is different from their models (although, as will be seen, it can account for the same dynamics that interested them).

While this book does not offer a mathematical model of culture per se, it shows how its account of meaning can be used to operationalize culture, understood as relatively shared interpretive grounds (plus the semiotic processes, qua meaningful public interactions and social relations, that are both mediated by and mediating of such grounds). It thereby resonates with the classic work of Urban on metaculture (2001) and the motion of culture (2010), however different its conceptual framework and mathematical formalism are. And it is thus quite distinct from other mathematical models of culture, such as those put forth by Boyd and Richerson (2005) and McElreath and Boyd (2008). That said, it is partially grounded in similar tools as these last two works: game theory, evolutionary dynamics, difference equations, fixed points, and the like. And it should appeal to similar kinds of readers: those wanting a naturalistic and mathematical understanding of the emergence of conventions, values, social relations, and signs.

In short, by incorporating key insights and critiquing particular commitments across this wide range of literature, this model provides a relatively seamless integration of otherwise disparate topics, methods, and paradigms. The chapters that follow will apply this model to a series of distinct case studies as a means to explore its entailments and assess its merits.

I AGENTS THAT THINK

2 Symptoms and Sickness

Overview

This chapter introduces a relatively simple and easily generalizable scenario: the relation between sickness, symptoms, diagnosis, and treatment. Insofar as it is the first scenario to be explored, and insofar as some readers may be unfamiliar with the mathematical machinery, we will walk through the calculations in detail. Section 2.1 sets up the interpretive ground for the scenario. Section 2.2 determines the posterior probabilities using the priors and likelihoods in conjunction with Bayes' theorem. Section 2.3 determines the best interpretant for each sign, given this ground. Section 2.4 finds the highest price that one should pay for the medication. Sections 2.5 and 2.6 compare the economic value of a sign with its information value. Finally, the conclusion derives a measure for the value of an agent's interpretive ground in light of the facts on the ground. Very loosely, it calculates the value of a worldview in light of the world. In a certain sense, then, this chapter is not just about the shared concerns of economics and semiotics—it is also about the somewhat strained relation between subjective commitments and objective facts.

2.1 Grounding the Scenario

An illness is going around, and a certain percentage of the population has it. If you are healthy, you can go to work as usual. If you are sick, however, you cannot go to work for a week, and so you lose your weekly income. Besides knowing the baseline rate (or prior probability of having the illness), there is a diagnostic test you can take (for a small price), which tells you whether you have a certain symptom or not. Having the symptom is correlated with having the illness, and not having the symptom is correlated with not having the illness. That said, the correlation is not perfect; so even if you take the test, you will not know the status of your health for sure. Finally, if you think that it is likely that you have the illness, there is a medication that you may take (also for a small price). While this medicine is quite effective if you are sick, it also has a small chance of making you ill (if you happen to be healthy).

As just described, but switching from the impersonal *you* to a generic person, such a scenario has the following ontology:

$$O = \{O_1, O_2\}$$
$$= \{\text{person is healthy, person is sick}\};$$
$$S = \{S_1, S_2\}$$
$$= \{\text{person does not evince symptom, person evinces symptom}\};$$
$$I = \{I_1, I_2\}$$
$$= \{\text{person does not take medication, person takes medication}\};$$
$$C = \{C_1, C_2\}$$
$$= \{\text{person stays (or becomes) healthy, person stays (or becomes) sick}\}.$$

As may be seen, the constituents of this ontology are really events, states, and actions. For example, a person is sick (O_2), a person does not take the medicine (I_1), and so forth. As may also be seen, in this framing, *not* evincing a symptom is as much a sign as evincing a symptom. Note, then, that the constituents of this ontology may be framed as propositions that may or may not be fulfilled in various circumstances, or worlds.

Given various commitments and values, the first task is to calculate the best action (I) given the available evidence (S). In so doing, a range of questions may be answered. For example, given the cost of the medicine, and the cost of the diagnostic test, in relation to the amount of income that someone stands to lose if they are sick, when is it worth taking the medication? Is it ever worth taking the medication without taking the diagnostic test? And what is the most money a person should be willing to pay for the test?

Just to be clear, the semiotic agent whose ground is being described may be the potentially sick person in question, a care provider, or even an automated device that helps the person or provider make such a decision. Indeed, while it is unlikely that a given individual might undertake such calculations, a care provider tasked with providing the best, most ethical, or at least most cost-effective solution for a population of people might be expected to select their semiotic strategies in such a fashion.

Rather than assign specific numbers to an agent's commitments and values, the problem will be set up using the following variables, such that the analysis will be as general as possible, given the ontology in question. First, there is the prior probability that the person in question is healthy or sick (given the baseline illness frequency in the population):

$$\begin{aligned} P(O) &= \begin{bmatrix} P(O_1) & P(O_2) \end{bmatrix} \\ &= \begin{bmatrix} p & 1-p \end{bmatrix}, \end{aligned} \tag{2.1}$$

where p is a variable that may range from 0 to 1 and represents the agent's degree of subjective certainty that a randomly selected person in the population is healthy (prior to any diagnostic test). Like most of the certainties examined in this chapter, while such a value might be based on relatively objective frequencies, or facts on the ground, it could also be grounded in hearsay, stereotypes, or intuition.

Second, there is the probability of a sign given an object, which may be represented as the following likelihood matrix (where the slash / should be read as 'conditioned on'):

$$\begin{aligned} P(S/O) &= \begin{bmatrix} P(S_1/O_1) & P(S_1/O_2) \\ P(S_2/O_1) & P(S_2/O_2) \end{bmatrix} \\ &= \begin{bmatrix} 1-a & b \\ a & 1-b \end{bmatrix}. \end{aligned} \tag{2.2}$$

As may be seen, the closer a is to 0, the more being asymptomatic (S_1) correlates with being healthy (O_1); and the closer a is to 1, the more being symptomatic (S_2) correlates with being healthy. Similarly, the closer b is to 0, the more being symptomatic (S_2) correlates with being sick (O_2); and the closer b is to 1, the more being asymptomatic (S_1) correlates with being sick. We will usually assume that $a < 1/2$ and $b < 1/2$, such that the symptom really does correlate with the illness (if only weakly so). Again, such values represent subjective certainties that may or may not be based on objective frequencies and/or scientific facts.

Third, there are the eventualities: the probability of each consequent (C), conditioned on both the action that the person undertakes (I) and the status of their health (O). These may be framed as follows (where $\wedge$ denotes the logical operator *and*):

$$P(C_1/I \wedge O) = \begin{bmatrix} P(C_1/I_1 \wedge O_1) & P(C_1/I_1 \wedge O_2) \\ P(C_1/I_2 \wedge O_1) & P(C_1/I_2 \wedge O_2) \end{bmatrix} = \begin{bmatrix} 1 & 0 \\ 1-q & r \end{bmatrix}, \tag{2.3a}$$

$$P(C_2/I \wedge O) = \begin{bmatrix} P(C_2/I_1 \wedge O_1) & P(C_2/I_1 \wedge O_2) \\ P(C_2/I_2 \wedge O_1) & P(C_2/I_2 \wedge O_2) \end{bmatrix} = \begin{bmatrix} 0 & 1 \\ q & 1-r \end{bmatrix}. \tag{2.3b}$$

As may be seen from these slices through the $2 \times 2 \times 2$ matrix, if the person is healthy (O_1) and they do not take the medication (I_1), then there is a probability of 1 that they stay healthy (C_1), and a probability of 0 that they become sick (C_2). However, if they take the medication (I_2), then there is a probability of q that they become sick and a probability of $1-q$ that they stay healthy. In this way, the magnitude of q scales with the risk of side effects.

Similarly, assuming the person is sick (O_2) and they do not take the medication, then there is a probability of 1 that they stay sick (and a probability of 0 that they become healthy). If they take the medication, however, there is a probability of r that they become healthy (and a probability of $1-r$ that they stay sick). In this way, the magnitude of r scales with the efficacy of the medicine.

We will assume that a week's salary for the person in question is equal to 1 (in some convenient choice of units). Insofar as a sick person cannot go to work for a week, the value of being healthy is 1 and the value of being sick is 0. It will also be assumed that the cost of the treatment (c) is some fraction of the person's weekly salary. (For the moment, we will assume that the diagnostic test is free.) The following matrices use these assumptions to represent the value associated with each consequent as a function of the health of the person and the action they undertake:

$$V(C_1 \wedge I \wedge O) = \begin{bmatrix} V(C_1 \wedge I_1 \wedge O_1) & V(C_1 \wedge I_1 \wedge O_2) \\ V(C_1 \wedge I_2 \wedge O_1) & V(C_1 \wedge I_2 \wedge O_2) \end{bmatrix} = \begin{bmatrix} 1 & 1 \\ 1-c & 1-c \end{bmatrix}, \tag{2.4a}$$

$$V(C_2 \wedge I \wedge O) = \begin{bmatrix} V(C_2 \wedge I_1 \wedge O_1) & V(C_2 \wedge I_1 \wedge O_2) \\ V(C_2 \wedge I_2 \wedge O_1) & V(C_2 \wedge I_2 \wedge O_2) \end{bmatrix} = \begin{bmatrix} 0 & 0 \\ 0-c & 0-c \end{bmatrix}. \tag{2.4b}$$

As may be seen by these slices through the $2 \times 2 \times 2$ matrix, in the context of the conjunction of events O_1 (person is healthy), I_1 (person does not take medication), and C_1 (person stays/becomes healthy), the value for the agent is 1. In the context of the conjunction of events O_1 (person is healthy), I_2 (person takes medication), and C_1 (person stays/becomes healthy), the value for the agent is $1-c$. And similarly, for other combinations.

Insofar as there are a lot of variables to keep track of in this scenario, it is worthwhile summarizing them in a single table:

Variable	Description	Value
p	Prior probability of being healthy	$0 \leq p \leq 1$
a	Probability of being symptomatic if healthy	$0 \leq a \leq 1/2$
b	Probability of being asymptomatic if sick	$0 \leq b \leq 1/2$
c	Cost of treatment	$0 \leq c \leq 1$
r	Probability medicine is efficacious	$0 \leq r \leq 1$
q	Probability of side effects	$0 \leq q \leq 1$

The first column shows each of the key variables. The second column describes the dimension that each variable is meant to represent. Finally, the third column specifies the range of values that such variables can take in the scenario in question.

The next two sections will use these grounds to analyze this scenario in conjunction with the main equation from chapter 1.

2.2 Finding the Posteriors

Using the likelihoods and priors, we now want to find $P(O/S)$, also known as the posterior probabilities. These represent the agent's subjective certainty that the person is healthy or sick *after* learning the results of the diagnostic, and hence after the agent has sensed the sign. To do this, we first calculate $P(S)$, the probability of receiving S_1 or S_2:

$$
\begin{aligned}
P(S_i) &= \sum_k P(S_i/O_k) \cdot P(O_k) \\
&= P(S_i/O_1) \cdot P(O_1) + P(S_i/O_2) \cdot P(O_2) .
\end{aligned}
\tag{2.5}
$$

Using the likelihoods and priors from the last section, we find

$$
\begin{aligned}
P(S) &= \begin{bmatrix} P(S_1) & P(S_2) \end{bmatrix} \\
&= \begin{bmatrix} (1-a) \cdot p + b \cdot (1-p) & a \cdot p + (1-b) \cdot (1-p) \end{bmatrix} .
\end{aligned}
\tag{2.6}
$$

This is the probability of receiving one sign or the other (regardless of the objects they stand for).

We may now calculate $P(O/S)$ using Bayes' theorem (recall equation 1.2), which expresses the posterior probabilities in terms of the prior probabilities and likelihoods:

$$
\begin{aligned}
P(O_j/S_i) &= \frac{P(S_i/O_j) \cdot P(O_j)}{\sum_k P(S_i/O_k) \cdot P(O_k)} \\
&= \frac{P(S_i/O_j) \cdot P(O_j)}{P(S_i)} .
\end{aligned}
\tag{2.7}
$$

Using the results of the previous calculation, we find that the posterior probabilities are as follows:

$$
\begin{aligned}
P(O/S) &= \begin{bmatrix} P(O_1/S_1) & P(O_1/S_2) \\ P(O_2/S_1) & P(O_2/S_2) \end{bmatrix} \\
&= \begin{bmatrix} \frac{(1-a) \cdot p}{(1-a) \cdot p + b \cdot (1-p)} & \frac{a \cdot p}{a \cdot p + (1-b) \cdot (1-p)} \\ \frac{b \cdot (1-p)}{(1-a) \cdot p + b \cdot (1-p)} & \frac{(1-b) \cdot (1-p)}{a \cdot p + (1-b) \cdot (1-p)} \end{bmatrix} .
\end{aligned}
\tag{2.8}
$$

The first row of this matrix expresses how likely it is that a person does not have the illness (O_1), conditioned on their not having the symptom (S_1) or their having the symptom (S_2). The second row expresses how likely it is that a person does have the illness (O_2), conditioned on their not having the symptom (S_1) or their having the symptom (S_2). As may be seen, each of the columns sums to 1.

To help readers visualize the dependencies in question, figure 2.1 plots one of the posterior probabilities, $P(O_2/S_2)$, and hence the probability that the person is sick, conditioned on their presenting the symptom, as a function of parameters a and b for two possible values of p.

As may be seen, for relatively small values of a and b (the back corner of the graphs), such that not having the illness strongly correlates with not presenting the symptom, and having the illness strongly correlates with presenting the

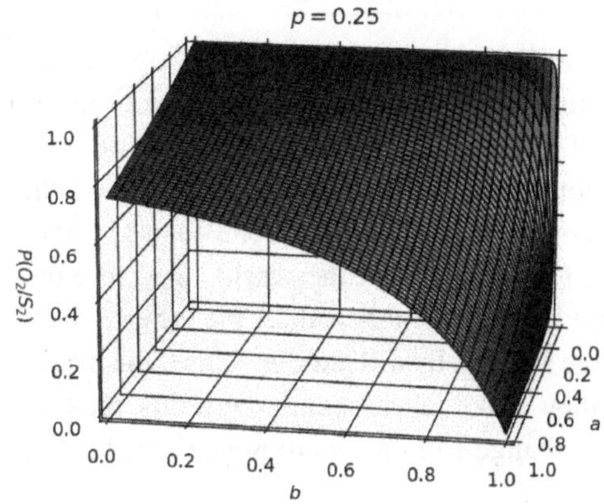

(a)

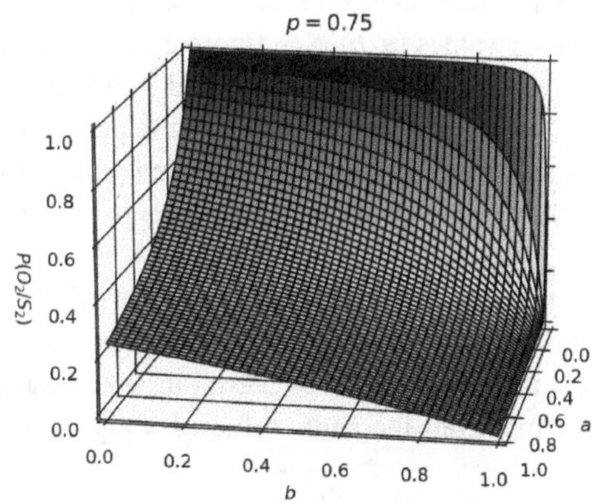

(b)

Figure 2.1: Posterior Probabilities

symptom, $P(O_2/S_2)$ is relatively high. This is especially true for small values of p (which, recall, is the baseline frequency of healthy people in the population such that $1-p$ is the baseline frequency of sick people in the population). As may also be seen, assuming a is very small, such that it is unlikely that a healthy person presents the symptom, the symptom provides good evidence for the illness, regardless of p or b. Similar plots could be offered for the other entries in $P(O/S)$.

To return to the discussion in chapter 1 regarding the meaning of a sign, understood as its semantic content or truth value, note the following: In one sense, the meaning of such signs is simply the types of objects, delimited in the ontology, that they are taken to correlate with: person is healthy (O_1) or person is sick (O_2). In another sense, it is the set of ontology-specific worlds in which such objects obtain: for example, all the worlds w_{ijkl} in which the first index (corresponding to O) is either 1 or 2, and all the other indices (corresponding to S, I, and C) are free to vary. In still another sense, the meaning of such signs is the change that they make to an agent's certainty regarding which object obtains, and thus the change from priors to posteriors, $P(O/S)-P(O)$, brought about by a sign.

The foregoing discussion highlights this last sense, which itself incorporates the first two senses. Were the top graph in figure 2.1 shifted down by .75, and the bottom graph shifted down by .25, corresponding to $P(O_2)=1-p$, they would show the *change* in the agent's certainty regarding the probability that the person is sick (O_2), conditioned on their evincing the symptom (S_2) as a function of the parameter values (a, b). See figure 2.2, which shows the meaning of S_2 with regard to O_2 as a function of a and b, when the agent has flat priors, such that $P(O_2)=P(O_1)=.5$. Any such point in parameter space thereby shows the direction (positive or negative) and the magnitude (how positive or negative) of the change in question, for a given interpretive agent and at a given time.

Such results are, it should be stressed, not particular to this scenario per se, but pertain to any similar set of sign-object relations as a function of Bayes' theorem. We will now use these results as partial inputs to our main equation, and thereby start moving from the truth value of a sign, so to speak, to its exchange value.

2.3 The Value of the Interpretants

We now want to find $V(I/S)$, the value of each interpretant given each sign, and hence the value of taking the medication (or not) conditioned, on whether a person presents the symptom (or not). To do this, we need to return to the main equation:

$$V(I_i/S_j)=\sum_k \sum_l V(C_k \wedge I_i \wedge O_l)\cdot P(C_k/I_i \wedge O_l)\cdot P(O_l/S_j)\,. \qquad (2.9)$$

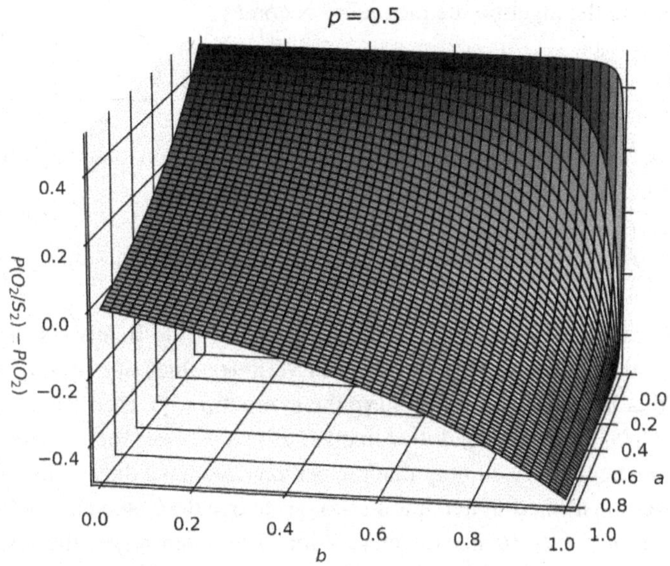

Figure 2.2: Meaning as the Change in Probability of an Object, Given the Sign

Suppose, for example, that an agent has sensed sign S_j, where $j = 1$ or $j = 2$. We want to know which of the following three cases is true:

- $V(I_1/S_j) > V(I_2/S_j)$, in which case the agent should instigate I_1,
- $V(I_1/S_j) < V(I_2/S_j)$, in which case the agent should instigate I_2,
- $V(I_1/S_j) = V(I_2/S_j)$, in which case the agent is indifferent.

In short, it is assumed that a rational agent that acts in accordance with its values and commitments, and hence in accordance with its interpretive ground, should (in the context of a particular sign) instigate the interpretant with the highest value (as conditioned on that sign).

Given the main equation, if $V(I_1/S_j) > V(I_2/S_j)$, it must be the case that

$$\sum_k \sum_l V(C_k \wedge I_1 \wedge O_l) \cdot P(C_k/I_1 \wedge O_l) \cdot P(O_l/S_j)$$

$$> \sum_k \sum_l V(C_k \wedge I_2 \wedge O_l) \cdot P(C_k/I_2 \wedge O_l) \cdot P(O_l/S_j). \quad (2.10)$$

In particular, if this inequality is true, then the agent should instigate I_1 in the context of sensing S_j. Putting in the values and certainties laid out in section 2.2, letting $j=1$ (meaning the person does *not* present the symptom), and undertaking the algebra, the inequality becomes

$$\frac{(1-a)\cdot p}{(1-a)\cdot p+b\cdot(1-p)}>\frac{(1-q-c)\cdot(1-a)\cdot p+(r-c)\cdot b\cdot(1-p)}{(1-a)\cdot p+b\cdot(1-p)}. \tag{2.11}$$

Similarly, but now letting $j=2$ (meaning that the person *does* present the symptom), the inequality becomes

$$\frac{a\cdot p}{a\cdot p+(1-b)\cdot(1-p)}>\frac{(1-q-c)\cdot a\cdot p+(r-c)\cdot(1-b)\cdot(1-p)}{a\cdot p+(1-b)\cdot(1-p)}. \tag{2.12}$$

To help readers visualize the relevant trade-offs, or relative sacrifices, figure 2.3 plots $V(I_i/S_j)$ as a function of p for various values of a, b, r, q, and c. Assuming that the person does not have the symptom (S_1), then wherever the solid line $V(I_1/S_1)$ is above the dotted line $V(I_2/S_1)$ it is better for the person not to take the medication. Assuming that the person does have the symptom (S_2), then wherever the dashed-dotted line $V(I_2/S_2)$ is above the dashed line $V(I_1/S_2)$, it is better for the person to take the medication. The points where the solid and dotted lines cross, or where the dashed-dotted and dashed lines cross, are thus *threshold values* where, given the certainties and values, a person would be indifferent insofar as both courses of action, I_1 and I_2, have the same value. In the two graphs on the lefthand side, $a=b=0.1$, such that not having the symptom is strongly correlated with not having the illness and having the symptom is strongly correlated with having the illness. In the two graphs on the right-hand side, $a=b=0.3$, such that while there is still a correlation, it is not as strong. As may be seen, the values for all interpretants for all signs increase as $P(O)=p$ increases, reflecting the fact that a person is less likely to be sick (given the baseline rates), such that they are less likely to lose a week's income. In all the graphs, we let $r=0.7$ and $q=0.2$, so the medication is relatively efficacious and the side effects are relatively small. Finally, while the medication costs 3/10 of a weekly salary in the top two graphs, it costs 1/10 of a weekly salary in the bottom-left graph and 1/2 of a weekly salary in the bottom-right graph. As may be seen, the value of taking the medication (I_2) either rises or falls with these shifts in price, while the value of not taking the medication (I_1) stays the same.

The next section will examine these critical prices, understood as points where $V(I_1/S)$ and $V(I_2/S)$ intersect, in more detail.

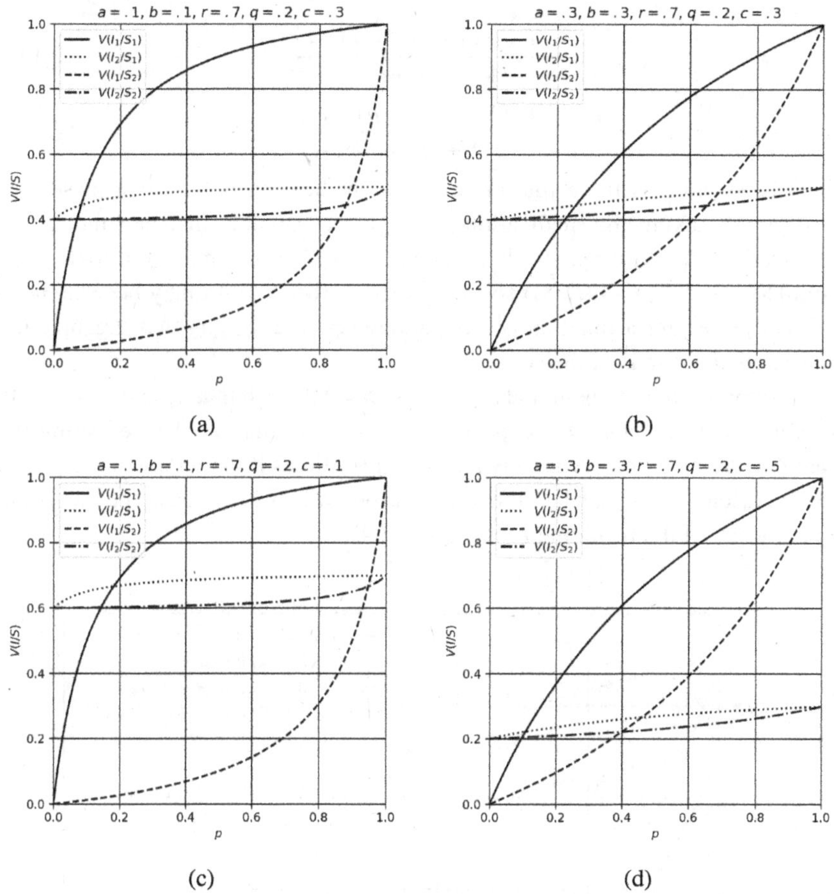

Figure 2.3: Conditional Values

2.4 Calculating the Critical Price

Solving for c in inequality (2.11), we find the highest price (c_1) that a person should be willing to pay for medication, assuming that they do not have the symptom (S_1). Solving for c in inequality (2.12), we find the highest price (c_2) that a person should be willing to pay for the medication, assuming that they do have the symptom (S_2). Doing the algebra, we find:

$$c_1 = \frac{r \cdot b \cdot (1-p) - q \cdot (1-a) \cdot p}{(1-a) \cdot p + b \cdot (1-p)}, \tag{2.13a}$$

$$c_2 = \frac{r \cdot (1-b) \cdot (1-p) - q \cdot a \cdot p}{a \cdot p + (1-b) \cdot (1-p)}. \tag{2.13b}$$

That is, so long as the actual price of the medication is less than these critical prices, taking the medication (I_2) is more valuable than not taking the medication (I_1). More precisely, $V(I_2/S_1) > V(I_1/S_1)$ so long as $c < c_1$ (for S_1). Similarly, $V(I_2/S_2) > V(I_1/S_2)$ so long as $c < c_2$ (for S_2). As may be seen, such critical prices are a function of the parameters (p, a, b, q, r) that organize the interpretive grounds of the scenario.

Assume for the moment that $a = 0.2$ and $b = 0.2$, such that S_1 and S_2 strongly correlate with O_1 and O_2, respectively. Assuming that $q = 0.2$, reflecting the fact that the medicine frequently leads to side effects if the recipient is healthy, these critical points may be plotted for various values of the baseline rate p and the efficacy of the medicine r (see figure 2.4).

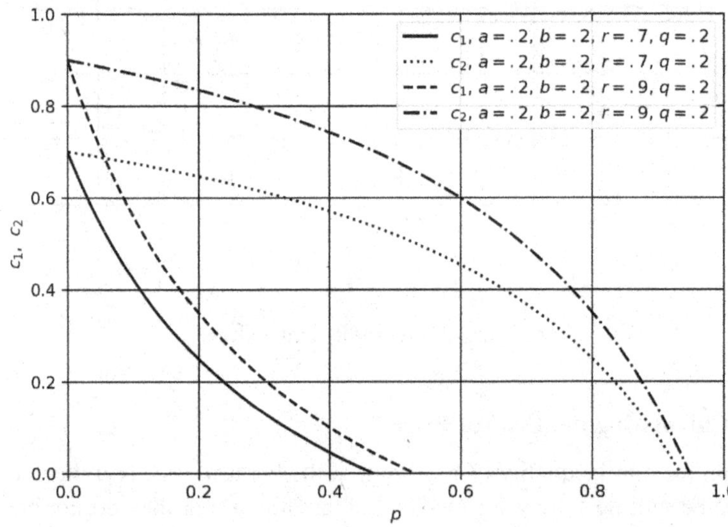

Figure 2.4: Critical Price Versus Prior Probability

Solid and dashed lines plot critical prices, as a function of $p = P(O_1)$, when a person does not present the symptom (S_1). Dotted and dash-dotted lines plot critical prices when a person does present the symptom (S_2). As may be seen, all four critical prices decrease as $p=P(O_1)$, the prior probability of a person's being healthy, increases. The price that a symptomatic person should be willing to pay (the top two lines) is higher than the price that a non-symptomatic person should be willing to pay (the bottom two lines). Finally, all else being equal, as the efficacy of the medicine (r) increases, the maximum price that a person should be willing to pay increases.

The next section will examine the worth of the diagnostic, loosely understood as the value of the information that it provides.

2.5 The Value of a Sign

Now suppose that the diagnostic test, which yields the sign S_1 (asymptomatic) or S_2 (symptomatic), costs a certain amount of money, which is a fraction of the person's weekly wage. They could make a decision about whether to take the medication based on their prior probabilities alone: $P(O)$. Or they could purchase the test, receive a sign, and update those priors: $P(O/S)$. Assuming that the diagnostic is relatively informative (an issue that we will return to in section 2.6), the information provided by the sign (regarding the status of their health, qua object) should help them make a better decision, such that the value of their expected outcome is greater than it otherwise would have been. (This is greater in value in the sense that, if they use the additional information that the sign provides to guide their actions, they are more likely to get better if ill, and less likely to fall ill if healthy). How might an agent assess this difference in value in a relatively precise way, such that they come to know how much the diagnostic test is really worth?

It is relatively straightforward to answer this question with the interpretive grounds already provided. First, suppose that a person does not pay for the diagnostic test (and hence does not know whether they present the symptom), so they make their decision based only on the information they already have. This is equivalent to determining the value of each action independent of the sign: $V(I)$. Such values may be calculated using the main equation, so long as we use prior probabilities, $P(O)$, instead of posterior probabilities, $P(O/S)$:

$$V(I_i)=\sum_k \sum_l V(C_k \wedge I_i \wedge O_l)\cdot P(C_k/I_i \wedge O_l)\cdot P(O_l)\,. \tag{2.14}$$

In particular, if $V(I_1)>V(I_2)$, then the person should instigate I_1 with an expected value $V(I_1)$. And if $V(I_2)>V(I_1)$, then the person should instigate I_2 with an expected value $V(I_2)$. Phrased another way, assuming that the person undertakes the action (I) with the highest value, the expected *prior* value of their action would be

$$\max\left(V(I_1), V(I_2)\right), \tag{2.15}$$

which may be calculated using the commitments and values from section 2.1.

Next, suppose that the person elects to undertake the diagnostic, so they receive a sign (S), which they can use to find their posteriors, $P(O/S)$, such that they can calculate $V(I/S)$ as in section 2.3. If they receive S_1, then they should instigate I_1 if $V(I_1/S_1) > V(I_2/S_1)$, and they should instigate I_2 if $V(I_2/S_1) > V(I_1/S_1)$, with an expected value of

$$\max\left(V(I_1/S_1), V(I_2/S_1)\right). \tag{2.16}$$

Similarly, if they receive S_2, then they should instigate I_1 if $V(I_1/S_2) > V(I_2/S_2)$, and instigate I_2 if $V(I_2/S_2) > V(I_1/S_2)$, with an expected value of

$$\max\left(V(I_1/S_2), V(I_2/S_2)\right). \tag{2.17}$$

Recall that $P(S_1)$ and $P(S_2)$, which were already calculated in section 2.2, are the probabilities of getting one sign or the other when taking the diagnostic. Assuming that an agent undertakes the most valuable interpretant (I_1 versus I_2) for a given sign (S_1 versus S_2), then the diagnostic allows them to guide their actions in such a way that their expected *posterior* value is

$$\sum_i P(S_i) \cdot \max\left(V(I_1/S_i), V(I_2/S_i)\right). \tag{2.18}$$

The difference between the expected posterior value and the expected prior value is the value of the sign, $\mathcal{V}(S)$. This is the maximum price that a person should be willing to pay for the diagnostic (all else being considered). In other words, and stated quite generally,

$$\mathcal{V}(S) = \sum_i P(S_i) \cdot \left[\max\left(V(I_1/S_i), V(I_2/S_i)\right) - \max\left(V(I_1), V(I_2)\right)\right]. \tag{2.19}$$

Figure 2.5 shows the way that $\mathcal{V}(S)$ scales with $p = P(O_1)$, as a function of various values and certainties. (Recall that the units of value being used are equivalent to the fraction of one's weekly salary.) As the correlation between the sign and object is increased, moving from $a = b = 0.3$ (the dotted line) to $a = b = 0.1$ (the solid line), the price that a person should be willing to pay (as some fraction of their weekly salary) increases, so long as the prior probabilities $P(O_1) = p$ are not too close to 0 or 1. This makes sense: the closer $P(O_1)$ is to 0.5, the more valuable the sign, insofar as the person is initially maximally ignorant about the state of their health. If we increase the efficacy of the medication (the dashed line) or increase the cost of the medication (the dashed-dotted line), but otherwise keep the value and certainties similar to what they are in the solid line, the value of the diagnostic test shifts accordingly.

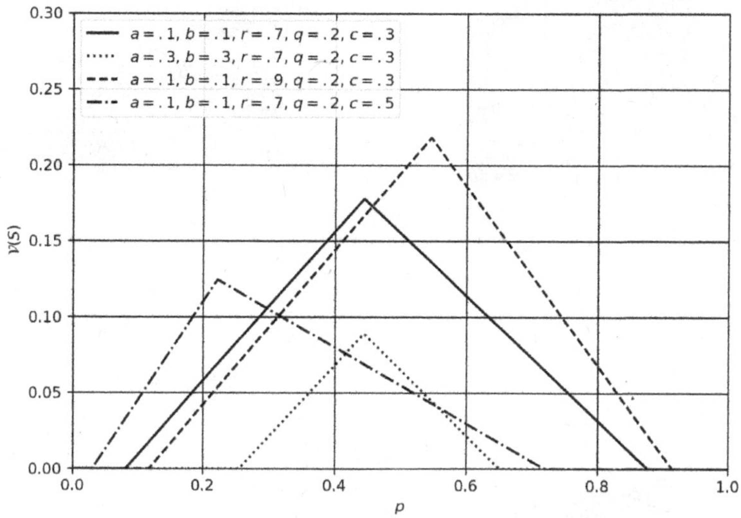

Figure 2.5: The Value of a Sign (in Two Dimensions)

Figure 2.6 shows the way that $\mathcal{V}(S)$ scales with a and b, for various values of p (when $r = 0.7$, $q = 0.2$, and $c = 0.3$). Horizontally flat regions of this graph correspond to those values of a and b where $\mathcal{V}(S) = 0$. As may be seen, and as should make sense, such regions most closely correspond to the bottom-right corner of the domain, where a and b are near 0.5, such that signs are minimally correlated with objects. Conversely, higher values of the graph correspond to regions where a and b are closer to 0. The closer $p = P(O_1)$ is to 0.5, the more valuable the sign, insofar as an agent's prior knowledge is minimally informative as to the status of the object. Finally, the points where the value of the sign is larger than zero all lie on a single plane, which shifts as a function of the other parameter values.

The next section will examine how $\mathcal{V}(S)$ compares with the information value of the sign, as calculated using a variety of classic measures. In effect, we will move from the exchange value of a sign to its surprise value.

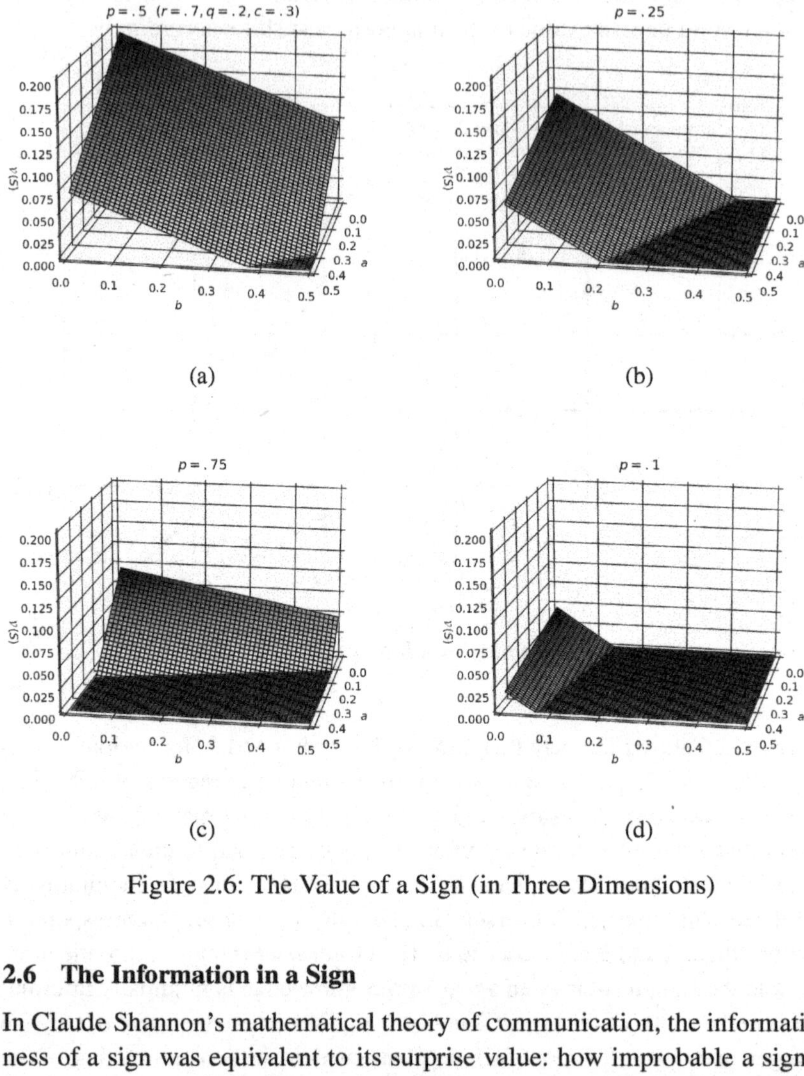

Figure 2.6: The Value of a Sign (in Three Dimensions)

2.6 The Information in a Sign

In Claude Shannon's mathematical theory of communication, the informativeness of a sign was equivalent to its surprise value: how improbable a sign is, given the relative probabilities of all the signs that might have been received instead. He took this to be the logarithm of the inverse probability of the sign:

$$H(S_i) = \log \frac{1}{P(S_i)} = -\log P(S_i)\,. \tag{2.20}$$

Building on this idea, he took the information capacity of a set of signs to be equivalent to the average surprise value of all the signs in the set:

$$H(S) = \sum_i P(S_i) \cdot \log \frac{1}{P(S_i)} = -\sum_i P(S_i) \cdot \log P(S_i). \qquad (2.21)$$

Famously, this way of quantifying information—which might be called "Shannon entropy"—is closely related to Boltzmann's definition of entropy. To some degree, this relation should accord with our intuitions. The more entropy, or disorder that a physical system has, insofar as its matter and energy are relatively dispersed, the less certain we are about the state it is in. Analogously, the more information capacity a set of signs has, the more surprising—and/or less predictable—any particular sign is. (For interested readers, Appendix A offers a detailed discussion of the relation between the entropy, information, value, and free energy.)

This account of information has long been critiqued as a potential theory of meaning, not least by Shannon himself, insofar as it makes no reference to the content of the message or, as we would say, to the object of the sign. To make up for this deficiency, many scholars use a related measure, known as the relative entropy, or Kullback–Leibler divergence. For our purposes, this is a measure of the gain in information or change in uncertainty (regarding the state of the object) that arises by acquiring a sign. Using our notation, it may be expressed as follows:

$$\mathcal{I}(S_j) = \sum_i P(O_i/S_j) \cdot \log \frac{P(O_i/S_j)}{P(O_i)}. \qquad (2.22)$$

As may be seen, this measure takes into account the relationship between the prior probabilities $P(O)$ and the posterior probabilities $P(O/S)$. Using Bayes' theorem, it may be easily calculated, so long as the priors and likelihoods are known. And if we want to know the informativeness of a system of signs regarding a system of objects, we simply average over all the signs in the system:

$$\mathcal{I}(S) = \sum_{ij} P(S_j) \cdot P(O_i/S_j) \cdot \log \frac{P(O_i/S_j)}{P(O_i)}. \qquad (2.23)$$

This is equivalent to measuring the average change in uncertainty regarding the state of the object that arises by acquiring a sign. Assuming base 2, such that the units of measurement are bits, this measure returns the average number of yes-or-no questions a sign saves one from having to ask to learn the identity of the object. Indeed, with a little bit of algebra, $\mathcal{I}(S)$ may be easily shown to be equivalent to the difference between the prior entropy (of the objects) and the average posterior entropy (of the objects conditioned on the signs), which is the formulation used in chapter 1:

$$\mathcal{I}(S) = \sum_j P(S_j) \cdot \left[H(O) - H(O/S_j) \right]. \tag{2.24}$$

Figure 2.7 graphs $\mathcal{I}(S)$ using the ontology, priors, and likelihoods of our symptoms and sickness scenario. In particular, it plots the average relative entropy as a function of a and b for different values of p. As may be seen, signs are more informative to us when our original uncertainty as to the state of the object is large ($p \approx .5$) and/or when particular signs are strongly correlated with particular objects (insofar as a and b are close to zero). As may also be seen, given how small the domain of objects is in our sickness and symptoms scenario, the most information a sign can provide is 1 bit, insofar as it can, at best, save us from asking one question: Is the person healthy?

It should be emphasized that as interesting as $\mathcal{I}(S)$ is, it is still a relatively inadequate measure for any model of meaning. This is because it only takes into account the priors and the likelihoods, and thus is only sensitive to the relation between the object and the sign. It does not take into account the action of the agent, qua interpretant, the relation that action has to the consequent, or the way that the consequent-interpretant relation depends on the state of the object. And hence it is insensitive to eventualities. Moreover, it also makes no reference to the value of the interpretant, given the sign in relation to the relation between the object and the consequent. It is thus insensitive to the interests and/or the identity of the agent. In short, the relative entropy, or Kullback–Leibler divergence, will return exactly the same answer regardless of the eventualities and values of the agent in question. It is thus a wholly inadequate measure of meaning.

The information provided by a sign is only valuable to an agent, rather, if it informs the agent's decisions and/or guides the agent's actions—decisions and actions whose outcomes have real stakes for that agent. In particular, even if the sign provides really good evidence of the state of the object, and thereby transforms the agent's knowledge of this state, that does not really matter to the agent if their actions are minimally efficacious, in the sense that the future state of the world (C) is indifferent to the agent's actions. Moreover, even if the future state of the world is sensitive to the agent's actions, if that state does not make a difference in value to the agent, then it hardly matters what the agent does, so it certainly does not matter whether the agent takes into account the sign or not. Our measure of the value of a sign, given in equation (2.19), in contrast, takes into account all these factors. In this sense, $\mathcal{V}(S)$ is a far more informative (and/or valuable) measure of a sign's value (and/or informativeness) than $\mathcal{I}(S)$. (All that said, a modified version of Shannon entropy does

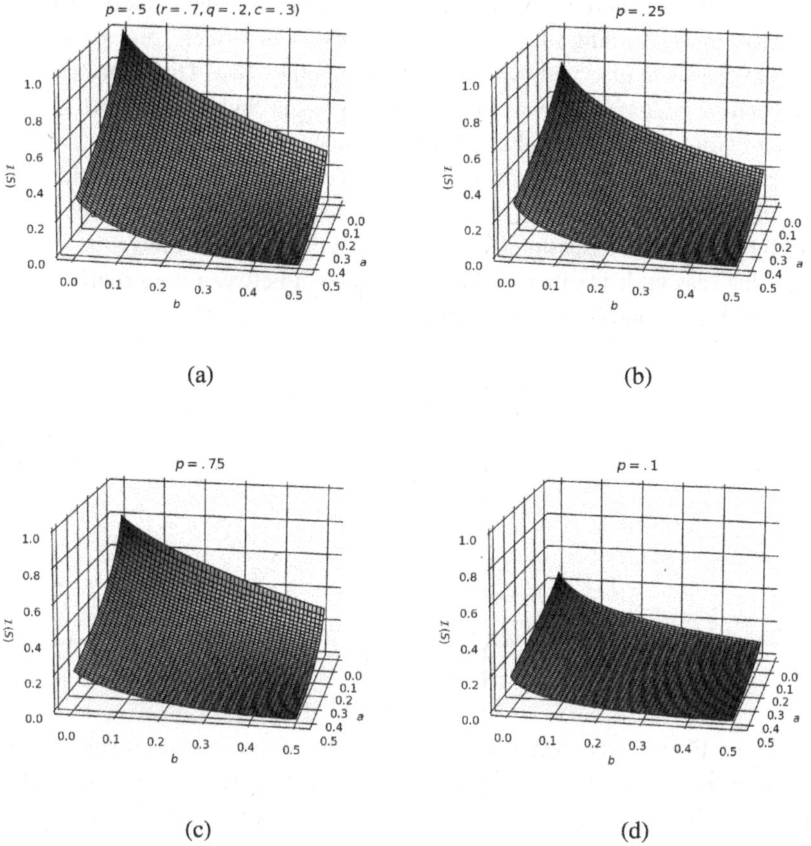

Figure 2.7: The Informativeness of a Sign

turn out to be useful as a measure of the agent's capacity to organize a universe by imposing constraints, as shown in Appendix B.)

The next section theorizes the difference in value between good grounds (which represent the environment in ways that are relatively accurate, undistorted, and precise) and bad grounds (which represent the environment in ways that are relatively inaccurate, distorted, or coarse). Insofar as an interpretive ground underlies all of an agent's decisions, a good ground—one that reliably captures the relevant patterns and resources in an environment, and thereby guides an agent's actions towards relatively useful outcomes—is one of the most valuable of metagoods. Indeed, it is arguably the arch-good.

2.7 Better and Worse Grounds

We have so far assumed that the interpretive grounds of agents—their relatively subjective understanding of symptoms, sickness, side effects, and so forth—are more or less in tune with the reality in which they live. This is because we have been interested in the ways that their inferences and actions were guided by their commitments and values. It should be stressed, however, that the representations that agents hold need not be in conformity with the worlds that they inhabit: they might have false beliefs about the relevant causal patterns, as well as erroneous assessments regarding possible risks and rewards.

For this reason, it is often useful to distinguish between the ground of the agent and the ground of the environment:

$$G_A = \begin{cases} P_A(O) \\ P_A(S/O) \\ P_A(C/I \wedge O) \\ V_A(C \wedge I \wedge O) \end{cases}, \tag{2.25a}$$

$$G_E = \begin{cases} P_E(O) \\ P_E(S/O) \\ P_E(C/I \wedge O) \\ V_E(C \wedge I \wedge O) \end{cases}. \tag{2.25b}$$

The priors, likelihoods, eventualities, and values of the former, as subscripted with an A, are relatively agent-specific and subjective. On the other hand, the priors, likelihoods, eventualities, and values of the latter, as subscripted with an E, are relatively agent-independent and objective. In other words, while G_A is properly called an "interpretive ground," G_E might be best described as "the facts on the ground" (at least so far as such facts are relevant to the scenario in question, as determined by the agent's interests, commitments, and capacities).

To return to the example of symptoms and sickness, it could be that the agent underestimates the baseline probability of the illness in the population, such that $P_A(O)$ is not in line with $P_E(O)$. Or it could be that the agent overestimates the efficacy of a medicine or underestimates the severity of the side effects, so that $P_A(C/I \wedge O)$ is not in line with $P_E(C/I \wedge O)$. Or it could be that the agent does not know the true cost of the treatment, so that $V_A(C \wedge I \wedge O)$ is not in line with $V_E(C \wedge I \wedge O)$, and so forth.

To be sure, as will be taken up in later chapters, just as G_A may reflect G_E (worldviews can come into alignment with the world, through experience, schooling, experiments, and the like), G_A may also destabilize G_E (worldviews can transform the world, typically through the practices of the agents that are committed to them, however unintentionally).

In short, the particular ground that an agent uses to make sense of an environment may be more or less resonant with the actual patterns and resources that constitute the environment—and dynamically so. In the best-case scenario, the ground of the agent may perfectly match the ground of the environment, such that the behavior of the agent is as valuable as it can possibly be (given the resources afforded by that environment and the capacities embodied by that agent). Conversely, in the worst-case scenario, the agent's ground might be deliberately warped—say, by the curse of warlock, or the greed of a media outlet—such that the agent instigates the least valuable, or most damaging, interpretant that it can in the context of the sign that it senses. It is more likely that there is some middle range, where the agent's ground is relatively motivated by, or reflective of, the environment in which it resides; such that the ground guides the agent's behavior towards relatively valuable outcomes. This is perhaps not the best outcome, but it is surely better than what would have occurred had the agent acted purely on chance or whim.

In light of the foregoing considerations, we are interested in calculating both the relative and absolute values of an agent's ground. In the case of absolute value, we mean how much the agent stands to gain or lose, on average, per sign it interprets, given the ground it has (insofar as that ground pushes the agent to instigate certain interpretants in the context of certain signs, which may have positive or negative repercussions for the agent, given the ground of the environment). In the case of relative value, we mean the difference between the absolute value, as just defined, and the maximum value that the agent could receive if its ground were perfectly consonant with, if not equal to, the ground of the environment.

Using G_A and G_E, as laid out earlier in this discussion, let us now turn to the measures in question. As usual, we will assume that the agent determines the relative value of particular interpretants in the context of particular signs using its ground, G_A, in conjunction with our main equation:

$$V_A(I_i/S_j) = \sum_{lk} V_A(C_l \wedge I_i \wedge O_k) \cdot P_A(C_l/I_i \wedge O_k) \cdot P_A(O_k/S_j)\,, \tag{2.26}$$

where the last term may be expanded using Bayes' theorem:

$$P_A(O_k/S_j) = \frac{P_A(S_j/O_k) \cdot P_A(O_k)}{\sum_m P_A(S_j/O_m) \cdot P_A(O_m)}\,. \tag{2.27}$$

Moreover, we will assume that the agent instigates the most valuable interpretant it can, given the sign that it senses, according to equation (2.26), so the agent's actual semiotic practices may be modeled as the softmax function of its conditional values:

$$P_A(I_i/S_j) = \text{softmax}\left(V_A(I_i/S_j)\right) = \frac{e^{\beta \cdot V_A(I_i/S_j)}}{\sum_m e^{\beta \cdot V_A(I_m/S_j)}}. \tag{2.28}$$

Using these ideas, the *absolute value* of the agent's ground, in the context of the environment's ground, may be defined as follows:

$$V(G_A/G_E) = \sum_{jkl} P_E(O_l) \cdot P_E(S_j/O_l) \cdot P_A(I_k/S_j) \cdot V_E(I_k/S_j). \tag{2.29}$$

In other words, to find the absolute value of the agent's ground, given the environment's ground, we sum over the product of the probability of an object (given G_E), the probability of a sign given an object (given G_E), the probability of an interpretant, given a sign (given G_A), and the value of such an interpretant in the context of such a sign (given G_E). Note that $V_E(I/S)$ is calculated in the same way that $V_A(I/S)$ is calculated, using equation (2.26) in conjunction with equation (2.27). The only difference is that the terms in G_E are used instead of the terms in G_A. As may be seen, whereas the first, second, and fourth terms on the right-hand side of equation (2.29) are relatively objective (turning, as they do, on the actual patterns found in the environment, via G_E), the third term is relatively subjective (turning, as it does, on the agent's projection of patterns onto the environment, via G_A).

To find the *relative value* of the agent's ground in the context of the environment's ground, we calculate the difference between the absolute value of the agent's ground and the absolute value of the environment's ground, where the latter is to be understood as the value that an agent would expect if its ground perfectly matched the actual ground of the environment, such that its behavior was as valuable as could be:

$$\begin{aligned} \Delta V_G &= V(G_E/G_E) - V(G_A/G_E) \\ &= \sum_{jkl} P_E(O_l) \cdot P_E(S_j/O_l) \cdot \left[P_E(I_k/S_j) - P_A(I_k/S_j)\right] \cdot V_E(I_k/S_j), \end{aligned} \tag{2.30}$$

where, by analogy with equation (2.28), $P_E(I_k/S_j) = \text{softmax}\left(V_E(I_k/S_j)\right)$. Insofar as the absolute value of the environment's ground will always be larger than, or equal to, the absolute value of the agent's ground, $\Delta V_G \geq 0$. The magnitude of ΔV_G thereby expresses the difference between the value that *could be* obtained by an agent and the value that is *actually* obtained—on average, and per sign interpreted.

Indeed, if an environment is such that it evinces no causal patterns and/or embodies no valuable resources, such that all interpretants are equally valuable in the context of any sign, then it is not only not worth it for an agent to have a good ground as opposed to a bad ground, it is also not worth it for an agent to have a semiotic ability in the first place. In this way, our measure not only tells us the relative value of good grounds and bad grounds, but also the relative value of semiosis per se. Framed in another way, if an environment is minimally organized, we should not expect there to be any agents that have evolved (learned or calculated) an interpretation of it.

It has so far been assumed that the ontology underlying the agent's ground is commensurate with the ontology underlying the environment's ground, in the sense that both ontologies make reference to identical sets of objects, signs, interpretants, and consequents. While a useful assumption, it is probably more useful to consider situations—like our own—in which the environment is far richer than most agents can imagine, such that the ontologies of agents (i.e., the types of objects and consequents that they can take into account and/or the types of signs and interpretants that they can sense and instigate) only scratch the surface of what is possible, given the patterns and resources in the environment per se. There could, for example, be other illnesses that the agent is not aware of, as well as other symptoms that they cannot sense, other treatments that they cannot instigate, and other consequents that they cannot fathom. How might we calculate the value of an agent's ground when the ontologies in question are incommensurate?

Here is a relatively simple way to do this. We calculate $P_A(I/S)$ as usual, using equation (2.28) in conjunction with equation (2.26) and equation (2.27), taking into account the agent's ground (qua commitments and values) and, in particular, the ontology that these refer to. And we calculate $V_E(I/S)$ as usual, via equation (2.26), replacing G_A with G_E, and the ontology that it refers to. (And we undertake both of these calculations even if the ontologies of the agent and environment are wildly incommensurate with each other.) In calculating $V(G_A/G_E)$, in contrast, while we sum over all objects in the environment's ontology, we only sum over signs and interpretants that are in the intersection of the agent's ontology and the environment's ontology, for those are the only entities and events that the agent can sense or instigate that are themselves causally resonant with the environment, and hence the only means (or media) through which the agent can be coupled to the environment.

In particular, if the agent cannot sense certain signs, then even if the environment is filled with objects that produce those signs, that agent will be indifferent to them. Conversely, if the agent can sense signs that are not produced by any objects in the environment, then the agent will never encounter such signs. Similarly, if the agent cannot instigate certain interpretants, then even if the environment is filled with objects that can be affected by those interpretants, that agent will never tap into those possibilities. Conversely, if the agent can instigate certain interpretants but the environment contains no objects that can be affected by such interpretants, that agent's actions will not yield any fruits or generate any causal entailments more generally.

More formally, we first rewrite equation (2.29) by summing over the objects in the environment:

$$\begin{aligned} V(G_A/G_E) &= \sum_{jkl} P_E(O_l) \cdot P_E(S_j/O_l) \cdot P_A(I_k/S_j) \cdot V_E(I_k/S_j) \\ &= \sum_{jk} P_E(S_j) \cdot P_A(I_k/S_j) \cdot V_E(I_k/S_j). \end{aligned} \tag{2.31}$$

But now, in carrying out the sums over the signs and interpretants, we only sum over signs and interpretants that are found in the ontologies of both the agent and the environment. The absolute value of the agent's ground in the context of the environment's ground may thus be expressed as

$$V(G_A/G_E) = \sum_{S \in S_E \cap S_A} \sum_{I \in I_E \cap I_A} P_E(S) \cdot P_A(I/S) \cdot V_E(I/S), \tag{2.32}$$

where S_E is the set of signs in the environment's ontology, S_A is the set of signs in the agent's ontology, I_E is the set of interpretants in the environment's ontology, and I_A is the set of interpretants in the agent's ontology.

Note, then, that if $I_E \cap I_A = \emptyset$ (meaning the intersection of the agent's set of interpretants and the environment's set of interpretants is empty), such that there is no overlap between what the agent (thinks it) can do and what can be done in an environment, then the value of the agent's ground is zero in that environment. Similarly, if $S_E \cap S_A = \emptyset$, such that there is no overlap between what the agent (thinks it) can sense and what can be sensed in an environment, then the value of the agent's ground is likewise zero in that environment. Finally, note that if the agent's ontology is commensurate with the environment's ontology, such that $S_E \cap S_A = S_E = S_A$ and $I_E \cap I_A = I_E = I_A$, equation (2.32) reduces to equation (2.29).

Once we have calculated the absolute value of an agent's ground when its ontology is incommensurate with that of an environment, we can calculate the relative value of the agent's ground in that environment using equation (2.30), which remains the same. In particular, the environment's ground is by definition commensurate with itself, such that $V(G_E/G_E)$ does not change.

Having worked through the details of a simple scenario involving a sickness and its symptoms, readers should see how similar analyses could be offered for an infinity of analogous scenarios—scenarios also involving uncertainty, information, risks, actions, and rewards. We could also dig down into the model of expected value that we have been using and adjust it to include expected utility, as well as to encompass the insights of relatively severe critiques of precisely such utility-based models (such as prospect theory, strong evaluation, and the like). We could thereby offer a microeconomics of semiotic processes and/or a meta-semiotics of economic practices. (Appendix C, for example, discusses the relation between value and utility in the context of this model.) We will leave it to readers to apply the model to other scenarios, and to tweak the model to include still other understandings of value.

The next chapter takes up a range of interrelated scenarios that turn on the continuous, as opposed to discrete, version of the main equation, and attempts to understand the real-time unfolding of semiotic interactions between two or more selecting agents—in particular, the dialogue that occurs, on interactional time scales, between predator and prey.

3 Predators and Prey

Overview

This chapter works through another scenario involving selecting agents: predators chasing prey in an open environment. As will be shown, the coupled world lines of such agents constitute a kind of conversation or dialogue. Unlike the scenario in chapter 2, which turned on the discrete version of the main equation (from chapter 1) as applied to a universe with 16 possible worlds, this scenario turns on the continuous version of the main equation as applied to a universe with an infinity of possible worlds. The first two sections set up the ground for this scenario, taking the perspectives of both the predator and the prey. Section 3.3 uses this ground to examine the temporal dynamics of predator-prey relations. Section 3.4 generalizes the ground, as well as its accompanying dynamics, for scenarios involving two spatial dimensions. And the final two sections analyze the dynamics of more complicated scenarios in which the agents interact not only with each other, but also with the surrounding terrain.

3.1 The Grounds of Predation

Some predatory agents, along with their prey, reside in an open environment. They can sense signs of the prey, infer their locations, and then instigate actions to better capture them. The prey are also agents, who can sense signs of the predators, infer their locations, and instigate actions to better elude them. In some sense, then, the prey constitute the salient environment of the predators and the predators constitute the salient environment of the prey. We are interested in the temporal dynamics of such organism-environment interactions: how predators and prey, in their attempts to capture and elude each other, move through space over time; and hence the interactional dynamics of such agents, as evinced in the dialogic coupling of their world lines.

This scenario has many moving parts, insofar as it turns on several agents interacting over time by means of iterated and coupled semiotic processes. To better understand its dynamics, we will start small and build up our model as we go. If we focus on a single semiotic process, undertaken by a single predator chasing a single prey in one dimension (such that both predator and prey are constrained to move along the x-axis, as it were), we have the following ontology:

$O = \{$prey is located at position O along the x-axis$\}$;
$S = \{$sign of prey is located at position S along the x-axis$\}$;
$I = \{$predator attempts to move to position I along the x-axis$\}$;
$C = \{$predator arrives at position C along the x-axis$\}$.

In other words, we think of the sign as an indexical trace of the object, such as the prey's freshest tracks, and what matters to the agent, qua predator, is the location of those tracks, insofar as they provide evidence of the prey's current position. The agent interprets these tracks by moving to a new position—a position that allows it to better capture the prey. As may be seen, we are overloading the notation, insofar as we are using terms like O and S to denote both variables and values. As may also be seen, the values of all the variables belong to one and the same set: the real numbers ($\mathbb{R}$). Any point along the x-axis, from negative infinity to positive infinity, is a potential place where tracks, predators, or prey may be. This is what was meant above by an "open environment."

Note, then, that in contrast to the previous scenario, in which each variable in the ontology could take one of two values, each variable in this ontology can take one of an infinity of values: any position along the x-axis. Insofar as there are four variables (O, S, I, C) at issue, the universe of an agent consists of all points in $\mathbb{R}^4$. In other words, we have moved from a discrete ontology admitting 16 possible worlds, to a continuous ontology where any point in a four-dimensional space constitutes a possible world. Indeed, as will be taken up later in this chapter, we can even bump up into the space of possible world lines, which is constituted by possible paths over time through the universe of two interacting agents, qua predator and prey. Such a universe, in which every "world" (or rather "world history") is a possible way that a prolonged interaction between predator and prey could have unfolded, is extremely large.

To handle such an ontology, we need to use the continuous version of the main equation:

$$V(I/S) = \int_{-\infty}^{\infty} \int_{-\infty}^{\infty} V(C \wedge I \wedge O) \cdot P(C/I \wedge O) \cdot P(O/S) \cdot dC \cdot dO\,, \qquad (3.1)$$

in conjunction with the continuous version of Bayes' equation:

$$P(O/S) = \frac{P(S/O) \cdot P(O)}{\int_{-\infty}^{\infty} P(S/O') \cdot P(O') \cdot dO'}. \tag{3.2}$$

As may be seen, while the summations have become integrals, the continuous version of our main equation still involves the priors, likelihoods, eventualities, and values of the agent's interpretive ground. Recall equations (1.1) and (1.2). We will discuss each of these items in turn.

Let us start with the simplifying assumption that the agent has flat priors. In other words, before sensing any sign, the agent assumes that the object is equally likely to be located anywhere along the *x*-axis between $-L/2$ and $L/2$, where L can be as large as we like. More formally, the probability that the object is found at position O is given by

$$P(O) = \begin{cases} \frac{1}{L} & |O| \leq L/2, \\ 0 & |O| > L/2. \end{cases} \tag{3.3}$$

We are interested in the limit as L goes to infinity, such that the object can be located anywhere along the *x*-axis (with vanishingly small probability). Loosely, this assumption about flat priors amounts to treating the agent as if it had a very shallow memory: when inferring the location of the object, it takes into account only the most recent sign it has sensed.

We will assume that the agent's likelihoods are shaped as a Gaussian distribution, or bell curve, such that

$$P(S/O) = \frac{1}{\sqrt{2\pi}v} e^{\frac{-(S-O)^2}{2v^2}}, \tag{3.4}$$

which is suitably normalized, such that

$$\int_{-\infty}^{\infty} P(S/O) \cdot dS = 1. \tag{3.5}$$

In other words, the agent assumes that the signs—qua indexical traces—that the object emits, or leaves behind, are normally distributed around the object's current position with a standard deviation, or width, equal to some adjustable parameter v. This parameter might be understood to scale with the acuity of the agent's vision and/or with the degree of coupling between the object's past movements and its present traces.

These likelihoods and priors, in conjunction with Bayes' equation, may be used to calculate the posteriors as follows:

$$
\begin{aligned}
P(O/S) &= \frac{P(S/O)\cdot P(O)}{\int P(S/O')\cdot P(O')\cdot dO'} \\
&= \lim_{L\to\infty} \frac{P(S/O)\cdot P(O)}{\int_{-L/2}^{L/2} P(S/O')\cdot P(O')\cdot dO'} \\
&= \lim_{L\to\infty} \frac{\frac{1}{\sqrt{2\pi}v} e^{\frac{-(S-O)^2}{2v^2}}\cdot 1/L}{\int_{-L/2}^{L/2} \frac{1}{\sqrt{2\pi}v} e^{\frac{-(S-O')^2}{2v^2}}\cdot 1/L\cdot dO'} \\
&= \frac{\frac{1}{\sqrt{2\pi}v} e^{\frac{-(S-O)^2}{2v^2}}}{\int_{-\infty}^{\infty} \frac{1}{\sqrt{2\pi}v} e^{\frac{-(S-O')^2}{2v^2}}\cdot dO'} \\
&= \frac{1}{\sqrt{2\pi}v} e^{\frac{-(S-O)^2}{2v^2}}.
\end{aligned}
\tag{3.6}
$$

As may be seen, in the simple case of flat priors, the posteriors are equal to the likelihoods. In other words, the probability of finding the prey at position O, conditioned on sensing the sign at position S, is shaped as a Gaussian distribution whose average is S and whose standard deviation is v.

We will assume that the eventualities are independent of O and are also shaped as a Gaussian distribution:

$$P(C/I\wedge O) = P(C/I) = \frac{1}{\sqrt{2\pi}m} e^{\frac{-(C-I)^2}{2m^2}}, \tag{3.7}$$

which is suitably normalized, such that

$$\int_{-\infty}^{\infty} P(C/I)\cdot dC = 1. \tag{3.8}$$

In other words, the consequent is independent of the object and is shaped as a Gaussian distribution whose average is I and whose standard deviation is m. The smaller the value of m, the sharper—or more precise—the movements of the predator. That is, if the predator attempts to go to I, it usually ends up at some position C that is very close to that location. (Recall that C and I can take on any number between negative infinity and infinity, corresponding to some position along the x-axis.) Like parameter v, parameter m may be understood as making reference to the body, or embodied capacities, of the agent—such as its acuity of vision, extent of reach, and precision of movement.

Finally, the values of the agent will be modeled using the following equation:

$$V(C\wedge I\wedge O) = N\cdot \frac{1}{\sqrt{2\pi}r} e^{\frac{-(C-O)^2}{2r^2}} - D\cdot \sqrt{(I-A)^2}. \tag{3.9}$$

As may be seen, this equation is a function not just of C, I and O, but also of A, understood as the position of the agent prior to moving. Just as the absolute value of $(C-O)$ is the distance between the agent and the object (after the agent's movement), the absolute value of $(I-A)$ is the distance that the agent endeavors to move (in order to capture the prey). The smaller the distance between C and O, the larger the value of the movement; the larger the distance between I and A, the smaller the value of the movement. In other words, it is assumed that the agent positively evaluates movements that will probably get it close to where it assumes the prey is; and it negatively evaluates movements that will involve traveling long distances. Its actual movement will thus reflect a trade-off between these costs and benefits, caloric or otherwise. In particular, variable D is just a parameter that expresses the cost, energy loss, or pain of moving a unit distance, as it were. Similarly, variable N is just a parameter that expresses the nutritional benefit, caloric input, or pleasure of capturing the prey (and/or moving near to it). Finally, r might be understood as a distance "within claws reach" of the predator: so long as the prey is within that distance from the predator, given the predator's position, the predator has a good chance of capturing the prey. (Note, in reference to the last term in this equation, that taking the square root of the square of a difference is equivalent to taking the absolute value of the difference.)

We choose these equations, qua commitments and values, for our model because they are complex enough to generate interesting behavior, scalable enough to allow for easy extensions to higher dimensions and multiple agents, and simple enough to allow an analytic solution to our equations. In particular, putting these priors, likelihoods, eventualities, and values into our main equation, we find

$$\begin{aligned} V_{pred}(I/A \wedge S) &= \int_{-\infty}^{\infty}\int_{-\infty}^{\infty} V(C \wedge I \wedge O)\cdot P(C/I \wedge O)\cdot P(O/S)\cdot dC\cdot dO \\ &= \int_{-\infty}^{\infty}\int_{-\infty}^{\infty} N\cdot\frac{e^{\frac{-(C-O)^2}{2r^2}}}{\sqrt{2\pi r}}\cdot\frac{e^{\frac{-(C-I)^2}{2m^2}}}{\sqrt{2\pi m}}\cdot\frac{e^{\frac{-(O-S)^2}{2v^2}}}{\sqrt{2\pi v}}\cdot dC\cdot dO \\ &\quad -\int_{-\infty}^{\infty}\int_{-\infty}^{\infty} D\cdot\sqrt{(I-A)^2}\cdot\frac{e^{\frac{-(C-I)^2}{2m^2}}}{\sqrt{2\pi m}}\cdot\frac{e^{\frac{-(O-S)^2}{2v^2}}}{\sqrt{2\pi v}}\cdot dC\cdot dO\,. \end{aligned} \tag{3.10}$$

These integrals are readily calculated (given the simplifying assumptions that went into them), yielding

$$V_{pred}(I/A \wedge S) = N\cdot\frac{e^{\frac{-(I-S)^2}{2r^2+2m^2+2v^2}}}{\sqrt{\pi(2r^2+2m^2+2v^2)}} - D\cdot\sqrt{(I-A)^2}\,. \tag{3.11}$$

This function expresses the value of every possible interpretant in the context of every possible sign, given every possible position of the agent and the values of all five parameters. As the abbreviation *pred* indicates, these are the conditional values for the predator (in the role of agent) attempting to capture the prey (in the role of object). We have yet to calculate the conditional values for the prey (in the role of agent) that is attempting to elude the predator (in the role of object).

As may be seen in figure 3.1, which plots this function for various values of the parameters, there are usually two local maxima in the conditional values of the agent. One maximum is located at the peak of the Gaussian (where $I=S$) and the other maximum is located at the peak of the absolute value function (where $I=A$). As D decreases, the slope of the graph leading away from A decreases. As N increases, the overall size of the Gaussian increases. And as m, v, and r increase, the height of the Gaussian decreases and the width of the Gaussian increases.

It should be emphasized that the position of the interpreting agent A (and hence a core participant in the semiotic event) is a key parameter in this value function. Such conditional values are thereby *shifters*, insofar as they are contextually dependent on the current position of the interpreting agent, and thereby change in value—or shift—as a function of that agent's position.

So long as $V_{pred} \geq 0$ when $I=S$, it is worthwhile for the agent to move to the peak of the Gaussian; otherwise, it should stay at the peak of the absolute value function. Phrased another way, implicit in this equation is the following critical threshold:

$$\theta_{crit} = \frac{N}{D} \cdot \frac{1}{\sqrt{\pi(2r^2 + 2m^2 + 2v^2)}}. \tag{3.12}$$

Assuming, as always, that the semiotic agent instigates the most valuable interpretant it can, given the sign it senses, then if the distance between S and A is smaller than this threshold, the best interpretant for the agent is S. However, if the distance between S and A is larger than this threshold, then the best interpretant for the agent is A. In other words, the agent should move to S if the movement is worthwhile; otherwise, it should stay where it is. As may be seen, in all but the top-right graph in figure 3.1, this distance is within the critical threshold—such that the agent will move away from its current position and attempt to move to the peak of the Gaussian.

Insofar as there are a lot of variables to keep track of, it is worthwhile summarizing them in a single table:

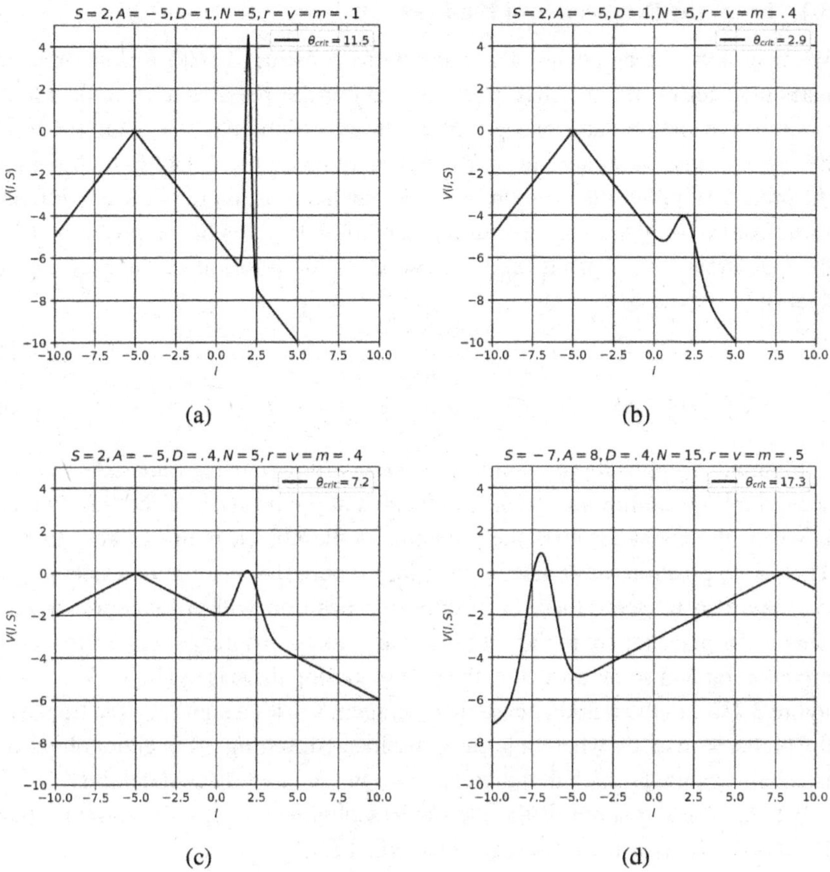

Figure 3.1: Predator's Evaluative Landscape (in One Dimension)

Variable	Description	Value
v	Standard deviation of $P(S/O)$	$0 \leq v$
m	Standard deviation of $P(C/I)$	$0 \leq m$
r	Standard deviation in $V(C \wedge I \wedge O)$	$0 \leq r$
D	Cost of movement	$0 \leq D$
N	Value of prey	$0 \leq N$

The first column shows each of the key variables. The second column describes the dimension that each variable is meant to represent. Finally, the third column specifies the range of values that such variables can take in the scenario.

Having derived the conditional values and critical distance for the predator, we will now do the same for the prey.

3.2 From the Perspective of the Prey

We may now use an analogous argument to construct *V*(*I*/*S*) for the prey. In particular, let us assume that everything about the prey, as a semiotic agent, is similar to the predator except that it places a high value on being far from the object rather than near it—for it strives to elude the object (which is now the predator), rather than capture it. Under such an assumption, the conditional values of the prey are identical to the conditional values of the predator, with the Gaussian term negated (and *A* now being the position of the prey as the agent):

$$V_{prey}(I/A \wedge S) = -N \cdot \frac{e^{\frac{-(I-S)^2}{2r^2+2m^2+2v^2}}}{\sqrt{\pi(2r^2+2m^2+2v^2)}} - D \cdot \sqrt{(I-A)^2}\,. \tag{3.13}$$

Figure 3.2 graphs this equation for a variety of parameter values. As may be seen, the local minimum is typically located at the position of the sign (which is where the agent assumes the predator, as the object, is most likely to be). That same position, of course, is the place where the agent least wants to go. As may also be seen, there is usually one position with a maximum value. This is the place where the agent most wants to go, insofar as it is far from the predator, but not so far away that the cost of getting there is too high. As seen in figure 3.2(b), in the unlikely case that the agent senses a sign (i.e., the tracks of the predator) exactly where it is currently located, setting off in either direction is equally valuable, such that there are two maxima with equal heights.

The critical threshold of the prey is less obvious. To find it, note first that $V_{prey}(I/A \wedge S)$ takes the following value when $I = A$:

$$V_{prey}(I/A \wedge S)|_{I=A} = -N \cdot \frac{e^{\frac{-(A-S)^2}{2r^2+2m^2+2v^2}}}{\sqrt{\pi(2r^2+2m^2+2v^2)}}\,. \tag{3.14}$$

Only if there is a point (*I*) where V_{prey} is higher than this value (or lower than the negation of this value) is it worthwhile for the prey to move. In other words, there must be some value of *I* where

$$N \cdot \frac{e^{\frac{-(I-S)^2}{2r^2+2m^2+2v^2}}}{\sqrt{\pi(2r^2+2m^2+2v^2)}} + D \cdot \sqrt{(I-A)^2} < N \cdot \frac{e^{\frac{-(A-S)^2}{2r^2+2m^2+2v^2}}}{\sqrt{\pi(2r^2+2m^2+2v^2)}}\,. \tag{3.15}$$

The points (*I*) where this relation holds are the points in each of the four figures that are worth moving to (given the parameter values in question). As may be seen, these are simply the points whose values are larger than the value of the point where the interpreting agent, as the prey, currently is. The point (or

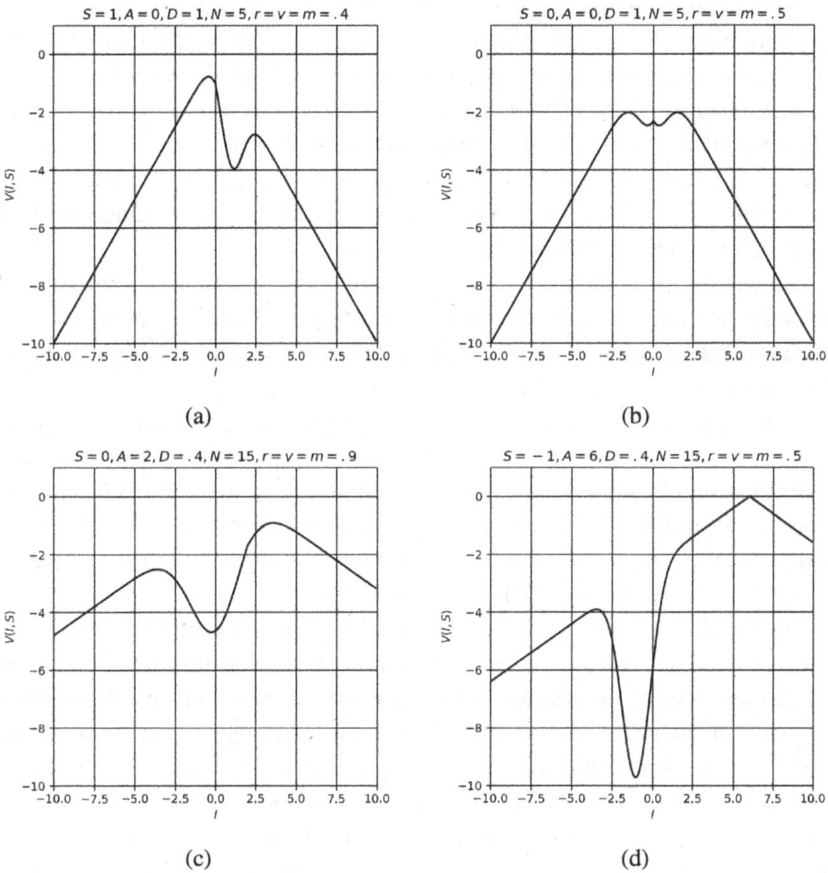

Figure 3.2: Prey's Evaluative Landscape (in One Dimension)

points, in symmetric contexts) in this region where V_{prey} is maximum is the point where the prey will attempt to move, as just discussed.

We now turn to the temporal dynamics of predator-prey interactions.

3.3 Predation as Conversation

We just derived the conditional values for predators and prey. These functions specify the value of an interpretant given a sign, in relation to the position of the interpreting agent, where the sign is a position along the x-axis (where the trace of the object is found), and the interpretant is a position along the x-axis (where the agent endeavors to move in order to best capture, or elude, the object in question—which happens to be the other agent).

We now assume that, having evaluated their next moves in such a way, the predator and prey attempt to move to the most valuable location (as given by their most currently calculated conditional values). Crucially, they do not move to that exact position. This is because the consequent (*C*) of their instigation (*I*) is not identical to their instigation, but rather probabalistically distributed around it in the shape of a Gaussian. Loosely, the effects of their actions never fully match their intentions. Recall, in particular, the assumption about eventualities that was laid out in equation (3.7).

We will let the computer choose a random number from such a distribution to determine the actual movement (*C*) of the agent. And we will assume that the agent's estimation of its precision of movement (*m*) in such an eventuality is equal to its actual precision.

(This is an easily adjusted presumption, and it is quite interesting to look at the agent's behavior when its own understanding of that behavior is at odds with its actual behavior, such that it overestimates or underestimates its own abilities. Recall the discussion at the end of chapter 2, when we examined the behavior of agents whose subjective grounds are at odds with the objective grounds of the environments they inhabit.)

Having thereby moved to better capture or elude the other agent (as the object), each agent leaves an indexical trace of its new position. This trace, itself the effect of the agent's most recent interpretant, will itself constitute the sign in the next semiotic process that the other agent will subsequently undertake, insofar as the trace-interpreting agent will again use it to infer where the trace-leaving agent is, such that it may determine where best to go next (and so on, and so forth). In short, not only does each semiotic agent have as its semiotic object the other agent, but the last interpretant of each such agent (or, rather, its consequent) helps constitute the next sign for the other agent.

Such traces will be distributed like a Gaussian around the position of the trace-leaving agent, as per the likelihoods described in equation (3.4). And we will again let the computer choose a random number from such a distribution to determine the position of the actual trace (*S*) left by each agent.

Letting unprimed variables denote the last time slice and primed variables denote the next time slice, and using O_1 and O_2 to denote the position of the predator and prey, respectively, the *interactional algorithm* looks as follows:

$$\begin{aligned} &...O_1 \xrightarrow{P(S/O)} S_1 \xrightarrow{V_{prey}(I/S)} I_2 \xrightarrow{P(C/I)} C_2 = O_2' \xrightarrow{P(S/O)} S_2'... \\ &...O_2 \xrightarrow{P(S/O)} S_2 \xrightarrow{V_{pred}(I/S)} I_1 \xrightarrow{P(C/I)} C_1 = O_1' \xrightarrow{P(S/O)} S_1'... \end{aligned} \tag{3.16}$$

And so it goes, with each agent using the trace of the other agent's last move to determine its next move, in relation to its current position, again and again.

Crucially, such an iterated semiotic process between two or more agents constitutes a kind of dialogue, where one agent's interpretant in the last semiotic process partially determines the other agent's sign in the current semiotic process. Phrased another way, if each agent is an organism, the other agent's behavior is the most salient (and labile) event in its environment. We may then track the position of each agent over time, across such iterations, and thereby follow its world line. In essence, then, our main equation allows us to calculate the temporal dynamics of complex interactions between two or more agents. This minimal coupling is what generates complex interactions, such that the dynamics of both agents together is greater than the sum of their individual dynamics.

Figure 3.3 plots some typical world lines for such two-agent scenarios. As may be seen, unless the predator is within the critical distance of the prey, its behavior is more or less equivalent to a random walk. Once it gets within the critical distance, it gives chase. The prey attempts to elude it until it finally gets out of the critical distance, in which case the behavior of both agents is again a random walk. (Not modeled here, obviously, is the behavior of the predator should it happen to capture the prey!) As may be seen, the initial positions were chosen such that the prey is near the critical distance of the predator, ensuring that we usually get coupled dynamics from the start.

Before continuing, let us offer a simple tweak of this scenario: when there are two prey and one predator. We assume that each of the two prey acts as before, but now the predator takes into account the position of both prey by taking into account each of their traces (denoted as S_1 and S_2). In particular, we assume that the predator's evaluative function simply averages over the possible positions of both prey such that its behavior is determined by the following equation:

$$V_{pred}(I/S_1 \wedge S_2) = \frac{N}{2} \cdot \left(\frac{e^{\frac{-(I-S_1)^2}{2r^2+2m^2+2v^2}} + e^{\frac{-(I-S_2)^2}{2r^2+2m^2+2v^2}}}{\sqrt{\pi(2r^2+2m^2+2v^2)}} \right) - D \cdot \sqrt{(I-A)^2}. \tag{3.17}$$

Figure 3.4 plots some typical world lines for such one-predator, two-prey scenarios. As may be seen, the results are similar to the last scenario, but show some interesting differences—such as when the agent leaves off chasing one prey to chase the other one (which it, presumably, determined it had a better chance of catching), or when the predator loses the prey, only to catch its tracks again a few moments later, such that it gives chase again.

As may also be seen, when a prey is far enough from a predator to feel no threat, such as the prey represented by the dotted line in figure 3.4(c), the

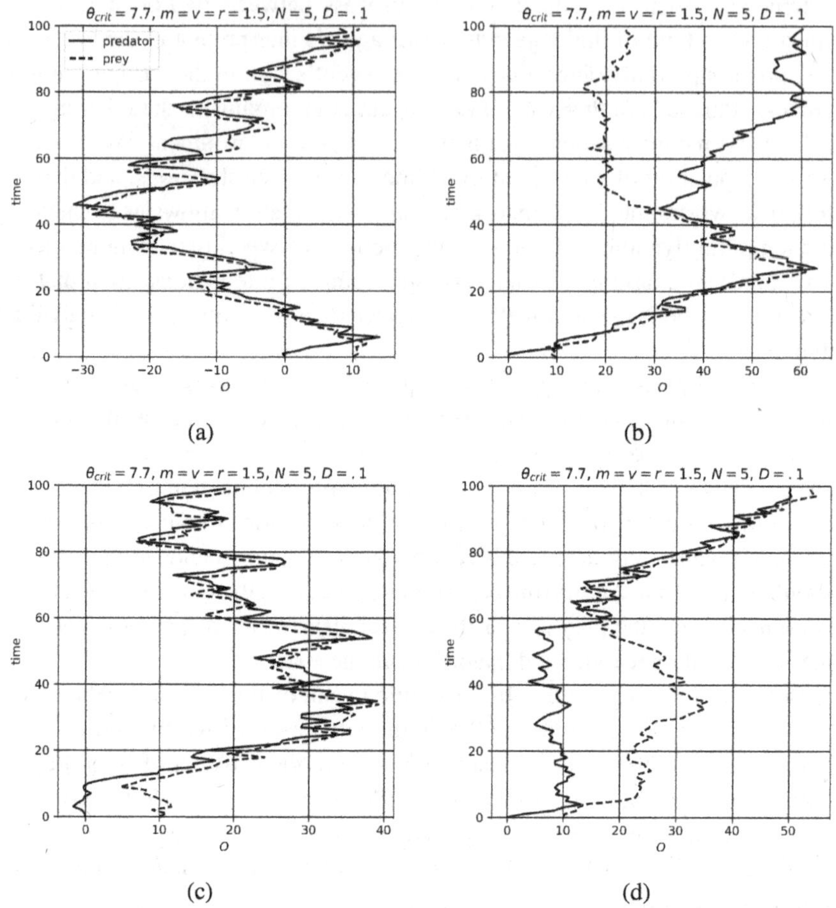

Figure 3.3: World Lines of Predator and Prey (in One Dimension)

prey engages in what is more or less a random walk, with a characteristic standard deviation. A similar fact holds for the predator. Only when the agents are within the critical thresholds of their objects do the dynamics of each agent become coupled, such that a qualitatively new kind of behavior emerges. This coupling can be stark and tight when the predator gets close to the prey and keeps on its trail, as evinced in the graphs wherever the solid line tracks a dotted or dashed line for relatively long periods of time.

The next section shows how the initial scenario can be easily generalized to two or more dimensions.

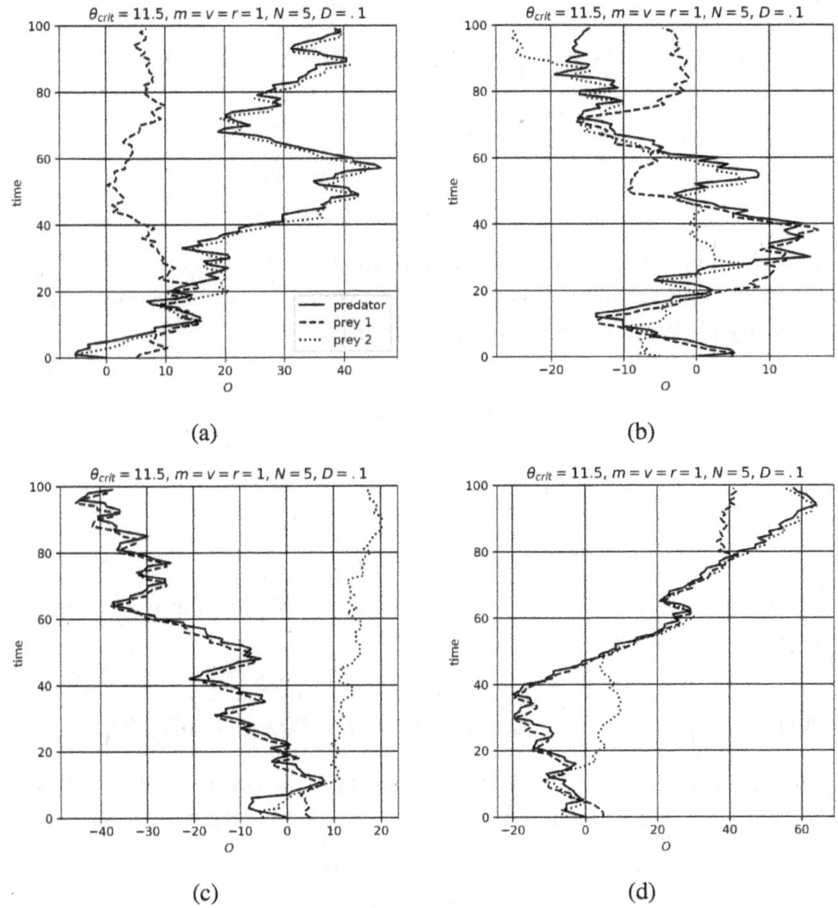

Figure 3.4: World Lines of One Predator and Two Prey (in One Dimension)

3.4 Two-Dimensional Dynamics

Suppose that the predator and prey now reside in an environment that is two-dimensional. The ontology of each agent now has eight variables instead of four: the position of the object, understood as a point in two dimensions (O_x, O_y), as well as the position of the sign (S_x, S_y), the position of the interpretant (I_x, I_y), and the position of the consequent (C_x, C_y). As before, the agent uses the position of the sign to infer the position of the object in order to determine how to best position itself to capture, or elude, the object (which is itself understood to be the other agent, and hence a closely coupled antagonist).

The main equation then becomes

$$V(I_x \wedge I_y / S_x \wedge S_y) = \iiiint V(C_x \wedge C_y \wedge I_x \wedge I_y \wedge O_x \wedge O_y) \cdot P(C_x \wedge C_y / I_x \wedge I_y \wedge O_x \wedge O_y) \cdot P(O_x \wedge O_y / S_x \wedge S_y) \cdot dC_x \cdot dC_y \cdot dO_x \cdot dO_y \,, \tag{3.18}$$

where the limits of integration for each of the four integrals are from negative infinity to positive infinity, as before. This equation can be expressed more compactly if the following conventions are used:

$$O_x \wedge O_y = O_{\hat{x}} \,, \tag{3.19a}$$

$$dO_x \cdot dO_y = dO_{\hat{x}} \,, \tag{3.19b}$$

$$S_x \wedge S_y = S_{\hat{x}} \,, \tag{3.19c}$$

$$dS_x \cdot dS_y = dS_{\hat{x}} \,. \tag{3.19d}$$

And it is similar for expressions involving I and C. Making these substitutions, the main equation thereby becomes

$$V(I_{\hat{x}} / S_{\hat{x}}) = \iint V(C_{\hat{x}} \wedge I_{\hat{x}} \wedge O_{\hat{x}}) \cdot P(C_{\hat{x}} / I_{\hat{x}} \wedge O_{\hat{x}}) \cdot P(O_{\hat{x}} / S_{\hat{x}}) \cdot dC_{\hat{x}} \cdot dO_{\hat{x}} \,. \tag{3.20}$$

The third-to-last term in this equation is simply the posteriors. Using Bayes' equation, these are a function of the priors and likelihoods in the usual way:

$$P(O_{\hat{x}} / S_{\hat{x}}) = \frac{P(S_{\hat{x}} / O_{\hat{x}}) \cdot P(O_{\hat{x}})}{\iint P(S_{\hat{x}} / O'_{\hat{x}}) \cdot P(O'_{\hat{x}}) \cdot dO'_{\hat{x}}} \,, \tag{3.21}$$

where the limits of integration for each of the two integrals in the denominator again range from negative infinity to positive infinity.

As before, the functions in question split up neatly and are readily evaluated. For these posteriors, we thereby get

$$P(O_{\hat{x}} / S_{\hat{x}}) = \frac{1}{\sqrt{2\pi v}} e^{\frac{-(O_x - S_x)^2}{2v^2}} \cdot \frac{1}{\sqrt{2\pi v}} e^{\frac{-(O_y - S_y)^2}{2v^2}} \,. \tag{3.22}$$

As may be seen, we are assuming that the standard deviation for vision (v) is the same along both dimensions.

We will continue to assume that the eventualities are independent of the position of the object, such that

$$P(C_{\hat{x}}/I_{\hat{x}} \wedge O_{\hat{x}}) = P(C_{\hat{x}}/I_{\hat{x}})$$
$$= \frac{1}{\sqrt{2\pi}m} e^{\frac{-(C_x - I_x)^2}{2m^2}} \cdot \frac{1}{\sqrt{2\pi}m} e^{\frac{-(C_y - I_y)^2}{2m^2}} . \tag{3.23}$$

As may be seen, we assume that the standard deviation for movement (m) is the same along both dimensions.

Finally, for the values, we have

$$V(C_{\hat{x}} \wedge I_{\hat{x}} \wedge O_{\hat{x}}) = N \frac{1}{\sqrt{2\pi}r} e^{\frac{-(C_x - O_x)^2}{2r^2}} \cdot \frac{1}{\sqrt{2\pi}r} e^{\frac{-(C_y - O_y)^2}{2r^2}} - D\sqrt{(I_x - A_x)^2 + (I_y - A_y)^2}, \tag{3.24}$$

where parameters r and N again denote something like the reach of the predator and the nutritional value of the prey, respectively; and where the cost function for movement is the appropriate two-dimensional generalization of the absolute value function (giving us something like an inverted an ice-cream cone).

Putting all these functions into the main equation and then evaluating the integrals, we find

$$V(I_{\hat{x}}/S_{\hat{x}}) = N \left(\frac{e^{\frac{-(S_x - I_x)^2 - (S_y - I_y)^2}{2v^2 + 2m^2 + 2r^2}}}{\pi(2v^2 + 2m^2 + 2r^2)} \right) - D\sqrt{(I_x - A_x)^2 + (I_y - A_y)^2} . \tag{3.25}$$

Equation (3.25) thereby describes the motivational landscape of the predator at any moment. An analogous set of steps would derive the equation describing the motivational landscape of the prey, which would differ only insofar as the first term on the right-hand side of this equation would be negated. While there were many steps involved in this derivation, each was the two-dimensional analog of a similar calculation undertaken in section 3.1. Readers should see the proper generalization to three dimensions as well.

As may be seen in figure 3.5, the predator (as the agent) is situated on top of the upside-down cone, and the prey (as the object) is situated at the top of the two-dimensional Gaussian. Insofar as the top of the Gaussian is higher than the top of the cone, the predator will interpret the landscape by moving there. As may be seen in figure 3.6, the prey (as the agent) is situated very near the predator (as the object), and so will endeavor to move to the highest point of the annular ridge—a point that, in effect, maximizes the prey's distance from the predator while minimizing the distance of travel.

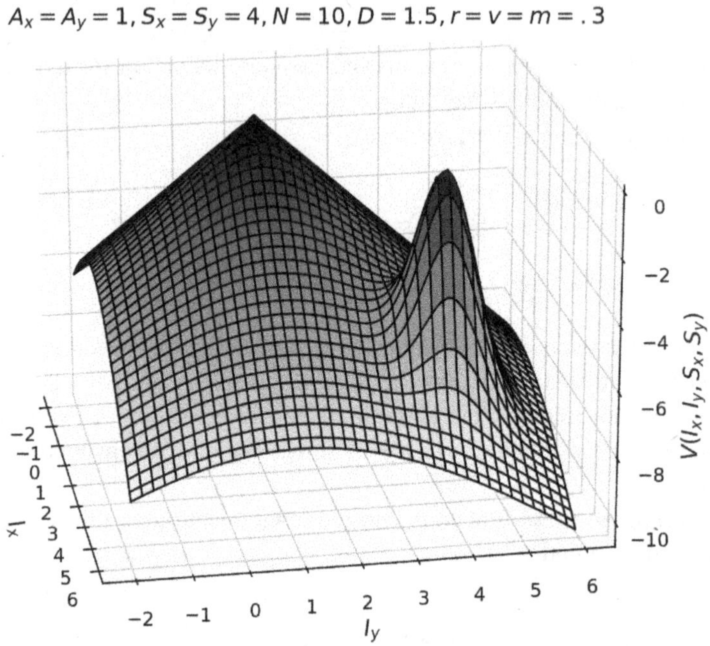

Figure 3.5: The Evaluative Landscape of a Predator (in Two Dimensions)

As in the one-dimensional scenario, we may now investigate the temporal dynamics of predator-prey interactions, and thereby track the world lines of such semiotic agents. Building on the analysis undertaken at the end of the last section, we will now assume that there are two predators chasing two prey. Figure 3.7 graphs the world lines of all four agents in such an encounter. As may be seen, after ganging up on a single prey, while ignoring the other, the predators split up, each tracking a different quarry. As interesting as these world lines are, we will not analyze the dynamics of this scenario any further, as they are similar enough to the one-dimensional scenario.

The next section will offer a useful generalization of our original scenario.

3.5 Landscape as Ally and Antagonist

Readers may have already imagined multiple variations of the foregoing scenarios. It is relatively straightforward, for example, to generalize to three

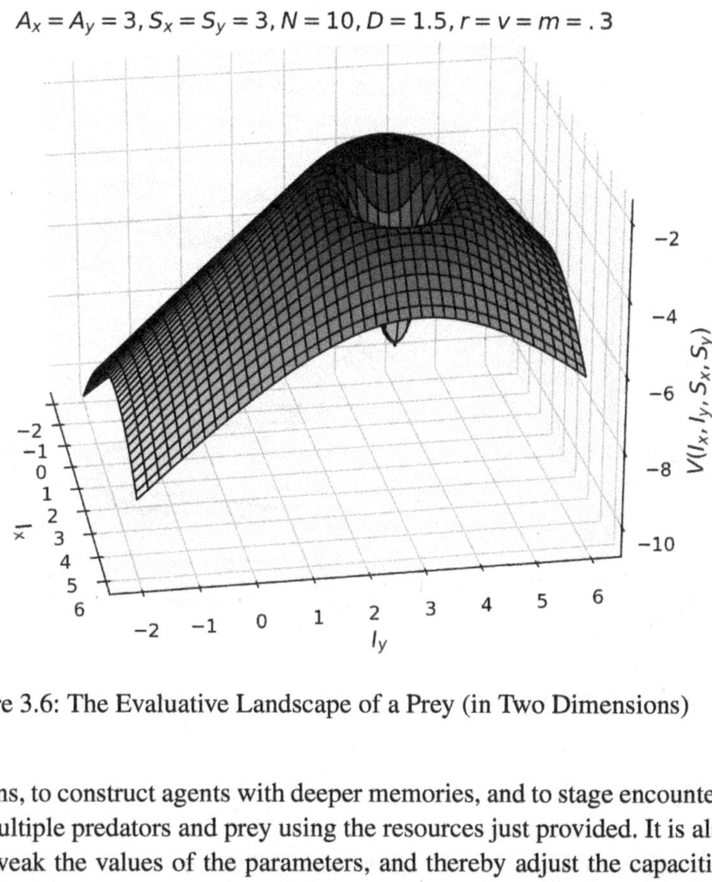

Figure 3.6: The Evaluative Landscape of a Prey (in Two Dimensions)

dimensions, to construct agents with deeper memories, and to stage encounters among multiple predators and prey using the resources just provided. It is also easy to tweak the values of the parameters, and thereby adjust the capacities and propensities of the agents, to analyze changes in their behavioral patterns. That said, we are not interested in working through all the variations of each scenario; we are interested, rather, in setting up some basic scenarios that readily admit analytic solutions with interesting and far-reaching dynamics. We will conclude, then, by focusing on a single variation: when the behavior of each agent not only takes into account the behavior of the other agent, but also the shape of the one-dimensional landscape in which their coupled interactions occur.

Suppose, in particular, that the landscape inhabited by the predator and the prey is not flat in all directions, but rather has some characteristic shape, qua changing altitude, as a function of position: $y = f(x)$. For example, these agents

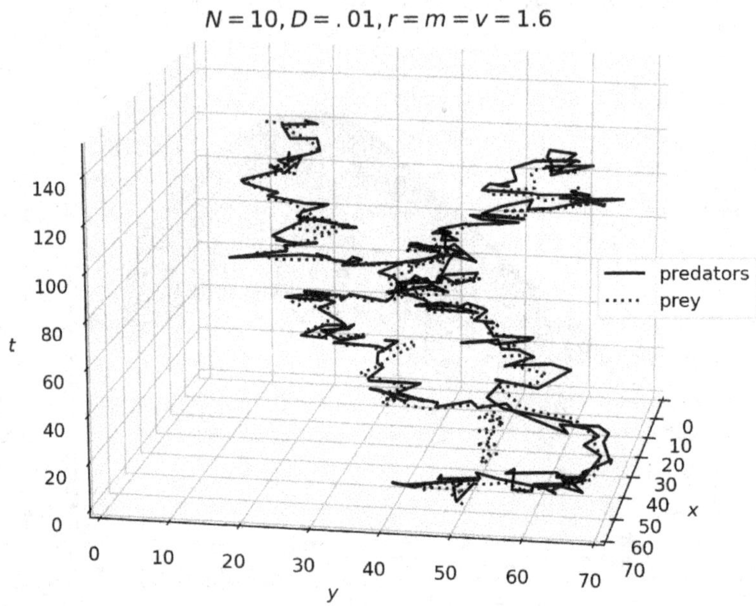

Figure 3.7: World Lines of Predator-Prey Interactions (in Two Dimensions)

might reside in a valley, where $y \approx x^2$; they might reside in rolling hills, where $y \approx \sin(x)$; and so forth.

Suppose further that both agents have to do work to overcome some kind of resistance whenever they move, due to the coupling of their bodies and the environment (e.g., mud and underbrush, or hooves and fur). We will assume that this frictional force is independent of their speed, as well the steepness of the terrain in which they move, such that the work that they must do in moving against it is proportional to the overall distance that they travel, regardless of the direction. More precisely, the work that they do (or the energy that they use) in moving from x_i to x_f is equal to some adjustable parameter F, which represents the force of friction, multiplied by the absolute value of the distance that they travel. In particular, so long as we know $y(x)$, we can calculate this distance, denoted as Δs, using the following integral:

$$\Delta s = \int_{x_i}^{x_f} ds = \int_{x_i}^{x_f} \sqrt{dx^2 + dy^2} = \int_{x_i}^{x_f} \sqrt{1 + \left(\frac{dy}{dx}\right)^2} dx. \tag{3.26}$$

Finally, suppose that the agents must not only work to overcome friction, they also work to move uphill (insofar as they struggle against gravity), and they have work done on them when they move downhill (insofar as gravity becomes their ally). More precisely, the work that they do (or have done on them) is equal to mg (the product of their mass and the acceleration of gravity) multiplied by their change in altitude, where the latter is simply

$$\Delta y = y(x_f) - y(x_i). \tag{3.27}$$

In such a situation, as should be clear, the shape of the landscape is a key player in the dynamics of predator-prey relations, insofar as both the length of an agent's path and the change in an agent's altitude depend on $y(x)$. To incorporate the role that the spatial environment plays, we can use the foregoing assumptions to model the energetic costs (or benefits) of moving in such a landscape. If we let ΔU denote the amount of energy used by an agent in moving from their initial position to some final position (and hence the amount of work that this movement requires), we find:

$$\Delta U = mg \cdot \left(y(x_f) - y(x_i)\right) + F \cdot \left| \int_{x_i}^{x_f} \sqrt{1 + \left(\frac{dy}{dx}\right)^2} \cdot dx \right|, \tag{3.28}$$

where we take the absolute value of the integral to ensure that the path length is always positive, no matter which direction the agent moves.

Insofar as ΔU gives the cost or benefit of moving from x_i to x_f, we will assume that the predator uses it to determine where best to move next. Equating A (the current position of the agent) with x_i and I (a potential place that the agent might endeavor to move) with x_f, we may construct a value function along the lines of equation (3.11):

$$\begin{aligned} V_{pred}(I/A \wedge S) = {} & N \cdot \frac{e^{\frac{-(I-S)^2}{2r^2+2m^2+2v^2}}}{\sqrt{\pi(2r^2+2m^2+2v^2)}} \\ & - mg \cdot \left(y(I) - y(A)\right) - F \cdot \left| \int_A^I \sqrt{1 + \left(\frac{dy}{dx}\right)^2} \cdot dx \right|. \end{aligned} \tag{3.29}$$

The first term on the right-hand side, as per our original ground, captures the fact that the predator strives to move as close as possible to S, understood as the place where it sensed the sign of the prey. And the second two terms are just the negation of ΔU, and hence represent the energetic cost of movement. (A similar equation will serve as the value function for the prey, so long as we add a negative sign in front of N to capture the fact that the prey strives to move away from the predator rather than towards it.)

As may be seen, in the limiting case of a perfectly flat environment where $y(x)$ is a constant function, such that $dy/dx = 0$, this equation simplifies to

$$V_{pred}(I/A \wedge S) = N \cdot \frac{e^{\frac{-(I-S)^2}{2r^2+2m^2+2v^2}}}{\sqrt{\pi(2r^2+2m^2+2v^2)}} - F \cdot |I-A|, \tag{3.30}$$

which is equivalent to equation (3.11), so long as we equate D with F, and remember that $|I-A| = \sqrt{(I-A)^2}$. Equation (3.29) thereby constitutes a far-reaching generalization of the original equation.

The next section will use this equation to model the behavior of our agents in various environments.

3.6 From Hyperbolic Valley to Rolling Hills

Suppose, for example, that the two agents reside in a valley such that $y(x) = \alpha \cosh(x/\alpha)$, where α is an adjustable parameter that loosely controls how wide the bottom of the valley is before it starts to rise, as well as the steepness of the valley's sides. In such a landscape, which might be deemed a hyperbolic valley, their cost function becomes

$$\begin{aligned}\Delta U &= mg \cdot \left(\alpha \cosh(I/\alpha) - \alpha \cosh(A/\alpha)\right) + F \cdot \left| \int_A^I \sqrt{1+\sinh^2(x/\alpha)}\, dx \right| \\ &= \alpha mg \cdot \left(\cosh(I/\alpha) - \cosh(A/\alpha)\right) + F \cdot \left| \int_A^I \cosh(x/\alpha)\, dx \right| \\ &= \alpha mg \cdot \left(\cosh(I/\alpha) - \cosh(A/\alpha)\right) + \alpha F \cdot \left| \sinh(I/\alpha) - \sinh(A/\alpha) \right|,\end{aligned} \tag{3.31}$$

where we have used the well-known facts that $d\cosh(x)/dx = \sinh(x)$ and $1 + \sinh^2(x) = \cosh^2(x)$. Such equalities make the integral easy to evaluate (such that we can get an analytic solution instead of having to numerically approximate it), which is one reason that this particular landscape was chosen in the first place.

Figure 3.8 shows some example runs of the behavior of predator and prey in just such an environment. All the parameters, including initial positions, have been held constant except for parameter α, whose value affects the relative steepness of the valley's walls. The dotted line in each graph shows the general shape of $y = \alpha \cosh(x/\alpha)$ for each parameter value. As α decreases in size, the valley's walls get steeper and steeper and, insofar as the predator and prey find it costly to climb uphill, they are pushed by the landscape to interact with each other in more tightly coupled ways. In effect, there is no escape, or even much rest, for the prey.

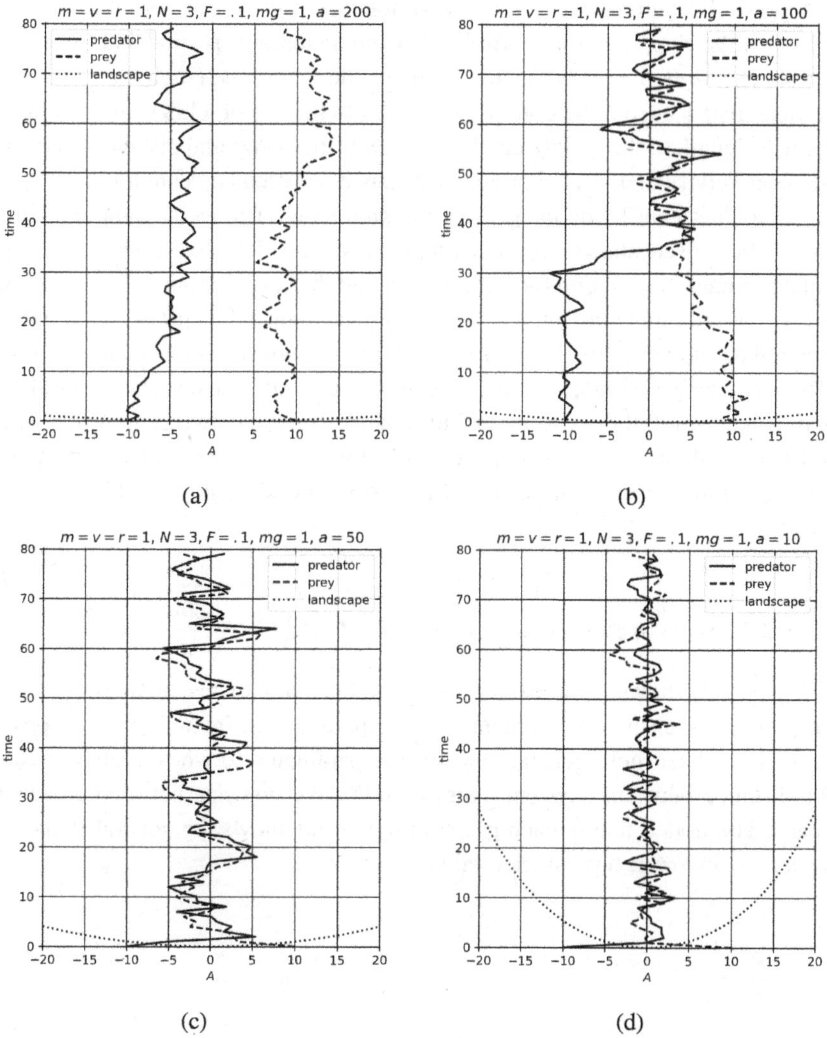

Figure 3.8: World Lines of Predator and Prey in a Hyperbolic Valley

Of course, from the standpoint of the analyst, it is now much easier to predict where the predator and prey are going to be. In particular, the steeper the walls of the valley, the less complex the behavior of the predator and prey, insofar as they will almost always be found near the bottom of the valley (with the exhausted prey probably already in the maw of the predator).

Hyperbolic valleys are relatively simple; slightly more complicated are rolling hills. Suppose, for example, that our landscape is modeled as $y(x) = a\cos(bx)$, where a adjusts the amplitude of the landscape (such that $2a$ is the vertical distance from the bottom of a valley to the top of a hill), and b adjusts the wavelength of the valleys (such that $2\pi/b$ is the horizontal distance between adjacent valleys and/or the horizontal distance between adjacent hills).

In such a scenario, the integral for the path length (given by equation 3.26) cannot be solved analytically but rather must be calculated numerically. To obtain an analytic solution for this integral, and thus also for our value function, we will approximate the integral by treating the sinusoidal curves as relatively triangular, and hence made up of upward-and downward-sloping lines, whose slope is given by $\pm 2ab/\pi$. (Understood as the vertical distance between a valley and its adjacent hill divided by the horizontal distance between the valley and hill.) In so doing, we find that the work done by the predator and prey, in moving from their current position (A) to some possible position (I), is given by

$$\Delta U = mg \cdot \left(a\cos(bI) - a\cos(bA)\right) + F \cdot |I - A| \sqrt{1 + \left(\frac{2ab}{\pi}\right)^2}. \qquad (3.32)$$

Figure 3.9 shows some examples of the behavior of the predator and the prey in just such an environment. All the parameters, including initial positions, have been held constant except for parameter b, whose value affects the distance between adjacent valleys, and/or the distance between adjacent peaks. The dotted line in each graph in figure 3.8 shows the general shape of $y = a\cos(bx)$ for each parameter value.

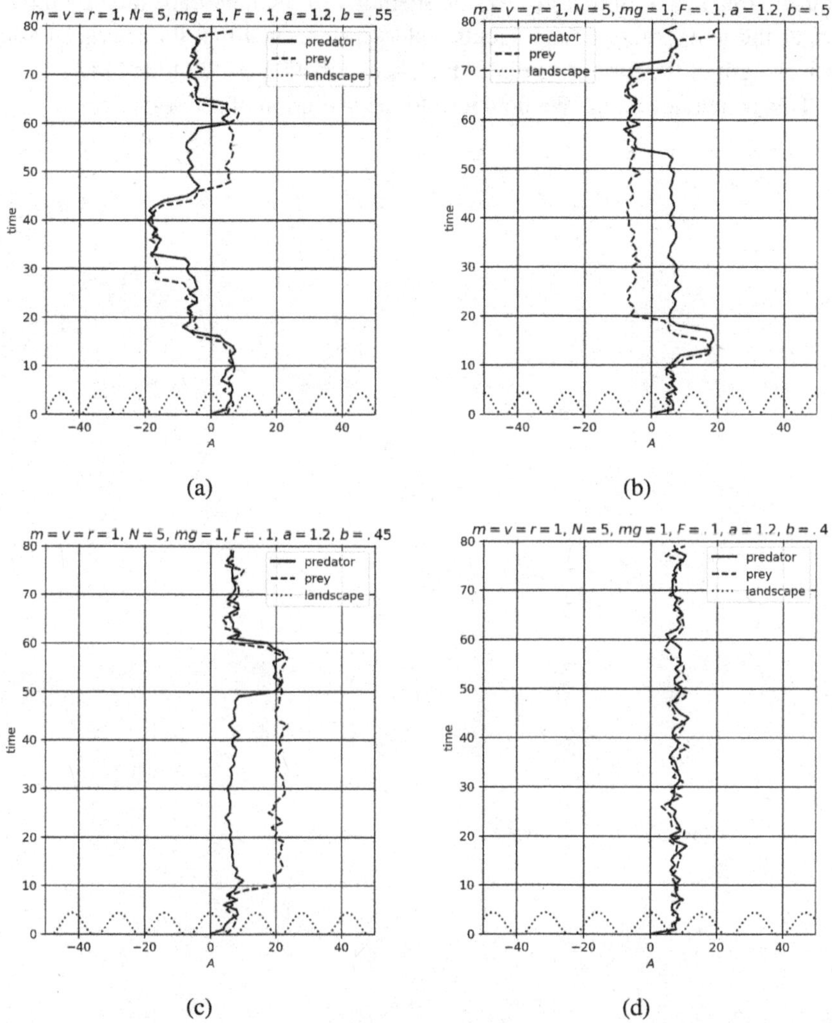

Figure 3.9: World Lines of Predator and Prey in Rolling Hills

As may be seen, the behavior of the predator and the prey is highly sensitive to the characteristic wavelength of the rolling landscape. If the distance between peaks is close enough, it is worthwhile for these agents to move across valleys. Nonetheless, they may get stuck in a particular valley for some stretch of time, either alone or with the other agent. As the distance between peaks increases, they stay isolated from each other, or coupled with each other, for longer stretches of time. Loosely, the spatiality of the landscape thereby influences the temporality of the agents' interactions, such that the characteristic wavelength of the former mediates the characteristic period of the latter.

This concludes Part I. We now turn to the evolution of semiotic agents.

II AGENTS THAT EVOLVE

4 Biosemiotic Agents

Overview

This chapter uses the main equation from chapter 1, in conjunction with classic ideas from mathematical biology, to model the evolutionary dynamics of biosemiotic populations. Section 4.1 introduces difference equations, fixed points, and phase spaces for simple kinds of biological systems. Section 4.2 reformulates the main equation so it can be used to calculate the relative fitness of alleles that mediate sign-interpretant relations. The next three sections use this reformulation, in conjunction with difference equations, to model the dynamics of an increasingly complex set of scenarios. Section 4.3 analyzes the evolution of a population of semiotic agents in a relatively fixed environment. Section 4.4 focuses on the coupled evolution of two populations, each of which constitutes a salient part of the other's environment. Finally, section 4.5 analyzes the relationship between the features of an emergent language and the affordances of the environment in which it emerges. The conclusion discusses the relation between significance and selection in biosemiotic contexts.

4.1 Evolution and the Difference Equation

Imagine an infinite population of relatively simple organisms in which there are n alleles for a particular gene: $A_1, A_2, \ldots, A_n$. Let $P(A_i)$ denote the proportion, or frequency, of each allele in the current generation, such that $\sum_{i=1}^{n} P(A_i) = 1$; and let $F(A_i)$ denote the relative fitness of each allele in the current generation. Assuming asexual inheritance, the frequency of each allele in the next generation may be calculated using the following difference equation:

$$P'(A_i) = \frac{P(A_i) \cdot F(A_i)}{\sum_j P(A_j) \cdot F(A_j)}. \tag{4.1}$$

In other words, the proportion of each allele in the next generation, denoted as $P'(A_i)$, is equal to its current proportion multiplied by its relative fitness, and then divided by the mean relative fitness of the population. Assuming that the relative fitness of all the alleles in each generation can be calculated, we can iteratively determine the changing frequencies of each allele across generations (beginning with some set of initial frequencies), and thereby model the evolution of the population over time.

In what follows, we will use the term *phase space* to denote an n-dimensional space, where n is the number of alleles in the population being modeled. Any point in such a space is a possible set of allele frequencies for the population in question, subject to the normalization constraint: $\sum_i P(A_i) = 1$. And the evolution of a population may be understood as a trajectory, or connected set of points, through such a space over time.

Of particular interest are *fixed points*, places in the phase space of the system where the frequency of each allele in the next generation is equal to its frequency in the current generation, such that the population stops evolving (unless it is perturbed). In other words, the population remains fixed at a particular position in phase space. Assuming a simple scenario in which there are only two alleles in the population, this occurs whenever $P'(A_1) = P(A_1)$ or, equivalently, $P'(A_2) = P(A_2)$. If we put either of these conditions, along with the normalization constraint for this system, $P(A_1) + P(A_2) = 1$, into equation (4.1), we find that it holds whenever

$$P(A_1) \cdot \big(1 - P(A_1)\big) \cdot \big(F(A_1) - F(A_2)\big) = 0 . \tag{4.2}$$

As may be seen, equation (4.2) is satisfied whenever $P(A_1) = 0$, $P(A_1) = 1$, or $F(A_1) = F(A_2)$. The first two conditions are easy to understand: a population that consists of only A_1 alleles, or only A_2 alleles, stays that way (unless a new allele is introduced through a mutation or the system is otherwise perturbed—say, by the influx of new members from a neighboring population).

The third condition is more complicated, insofar as the fitnesses, $F(A_1)$ and $F(A_2)$, may depend on the frequencies of the alleles in the population, $P(A_1)$ and $P(A_2)$, which may themselves be changing over generational time. Assuming for the moment that fitness is independent of frequency, then whenever $F(A_1) = F(A_2)$, every possible value of $P(A_1)$, from 0 to 1, is a stable fixed point. In other words, the alleles will remain at their initial frequencies in the population, whatever those happen to be, forever. When $F(A_1) > F(A_2)$, $P(A_1) = 0$ is an unstable fixed point and $P(A_1) = 1$ is a stable fixed point. In other words, so long as the initial frequency of A_1 is not equal to zero, organisms carrying this allele will take over the population. Finally, when $F(A_1) < F(A_2)$, $P(A_1) = 0$ is a stable fixed point and $P(A_1) = 1$ is an unstable fixed point. In other words, so long as the initial frequency of A_2 is not zero, organisms carrying this allele will take over the population.

Just to be clear, a population at an unstable fixed point will remain there indefinitely unless it is perturbed, meaning that even the slightest change in its frequency (due to a chance mutation, say, or new migrants) will cause it to move away from that point. Think of a ball at the top of a hill. A population at a stable fixed point, in contrast, will not only remain there indefinitely but, if perturbed by similar kinds of events, will move back to that point. Think of a ball at the bottom of a valley.

In short, the tendency of such populations is to move towards stable fixed points in phase space. The rest of this chapter will analyze the number, location, and stability of fixed points for populations of organisms whose alleles mediate various modes of signification and interpretation. Of particular interest is how such fixed points change as the values of the parameters are altered, insofar as such parameters mediate the semiotic abilities of organisms bearing various alleles.

4.2 The Relative Fitness of Biosemiotic Agents

To use difference equations and their fixed points to understand biosemiotic processes, we need to be able to calculate the relative fitness of the alleles that mediate those processes. In the first scenario examined in this chapter, this mediation is very simple: alleles (A_n) cause the organisms that carry them to be more or less likely to trigger certain events in the context of other events. Looking slightly ahead, the triggering events will be called *signs*, and the triggered events *interpretants*; and such mediation will be represented as follows:

$$A_n \rightarrow P_{A_n}(I_i/S_j)\,. \tag{4.3}$$

In other words, allele A_n causes the organism that carries it to cause I_i in the context of S_j with conditional probability $P_{A_n}(I_i/S_j)$. Such allele-specific conditional probabilities are thus the behavioral phenotypes, or *semiotic strategies*, of such organisms.

To motivate the foregoing choice of terms, it is assumed that the environment (which may include other organisms, or even other parts of the same organism) is organized in such a way that the signs are correlated with certain *objects* (insofar as they reliably follow from them), and the interpretants are correlated with certain *consequents* (insofar as they reliably lead to them). It is also assumed that the relative fitness conferred upon an organism by doing I in the context of C and O is $V(C \wedge I \wedge O)$. If all these assumptions are put together, the expected fitness of an organism bearing allele A_n is

$$F(A_n)=\sum_{ijkl} P(O_i)\cdot P(S_j/O_i)\cdot P_{A_n}(I_k/S_j)\cdot P(C_l/I_k\wedge O_i)\cdot V(C_l\wedge I_k\wedge O_i)\,. \quad (4.4)$$

This equation may be expressed more compactly. If $P(S_j)$ denotes the probability that the organism encounters sign S_j during a generation and $V(I_k/S_j)$ denotes the relative fitness conferred upon the organism by causing interpretant I_k in the context of sign S_j, then the expected relative fitness of an organism bearing allele A_n is simply

$$F(A_n)=\sum_{jk} P(S_j)\cdot P_{A_n}(I_k/S_j)\cdot V(I_k/S_j)\,, \quad (4.5)$$

where the last term on the right-hand side is specified by our main equation, as was introduced in chapter 1:

$$V(I_k/S_j)=\sum_{il} V(C_l\wedge I_k\wedge O_i)\cdot P(C_l/I_k\wedge O_i)\cdot P(O_i/S_j)\,. \quad (4.6)$$

Note the important shift that has occurred. Until now, we have been studying relatively mindful agents that use the main equation—or at least act as if they are using something like it—to determine which interpretant is most valuable given a sign (insofar as these two components are, or at least seem to be, causally related to objects and consequents in a certain manner). Such agents are compelled by rationality, or some other regimenting mechanism that promotes relatively rational behavior (if only their doctor's advice), to instigate the interpretant that will maximize their expected value in the context of the sign (given their scheme of evaluation and their assumptions about causality, however incorrect or unconscious). We are now studying relatively mindless agents that flourish or founder depending on whether their alleles push them to semiotically interact with a causally-patterned and resource-rich environment in a way that confers more or less fitness upon them. To return to the concerns raised at the end of chapter 2, the interpretive ground of the organism, as embodied in its alleles, comes to reflect the facts on the ground in the environment. For these reasons, both the allele and any organism bearing the allele will be called an *agent* in what follows. And we will often shift from speaking about the fitness of an allele to speaking about the fitness of an agent bearing the allele, given the relative simplicity of the systems that are analyzed.

Such organisms do not even necessarily sense S and instigate I; rather, their bodies and environments are organized in such a way that the presence of S (which itself is triggered by O) can trigger I (which itself triggers C). Insofar as an allele confers relatively more fitness on the organism that bears it (given the relation between C and O, in relation to the organism and its interests, in a particular environment), the frequency of that allele will increase in the population such that the population itself will become semiotically adapted to its environment. In effect, the organisms in the population end up behaving in a relatively rational manner, given the alleles at their disposal.

To be sure, and as will be shown in the sections that follow, what is a good interpretant of a sign given one set of allele frequencies (and hence at one position in phase space) may be a bad interpretant of the sign given another set of allele frequencies; and the behavior of such organisms may have long-term consequences on the environment that undercut the environmental patterning that generated them, such that their behavior can be quite complicated and their relative success may be short-lived, if not paradoxical. As will be discussed in the conclusion of this chapter, this makes it difficult to say what the meaning of a sign is or from whence comes the motivation for it unless the population comes to some kind of equilibrium. Nonetheless, we will sometimes speak of the meaning (value or information) of a sign that is out of, yet en route to, equilibrium—however short-lived.

To familiarize readers with this framework, the next two sections will carry out an extended analysis of a relatively simple scenario. As will be seen, equation (4.4), in conjunction with equation (4.1), give us all we need to begin our analysis of the evolutionary dynamics of biosemiotic systems.

4.3 Dynamic Organism, Fixed Environment

Suppose that there is an environment with two kinds of plants, which we may label O_1 and O_2. Whereas O_1 plants have smooth leaves (S_1), O_2 plants have wrinkled leaves (S_2). Not only are such plants more or less abundant in the environment, as specified by their relative probabilities, $P(O)$, they also have different nutritional values, as will be specified by the fitness-promoting value that they confer on the agent that consumes them.

Suppose that a population of plant-eating animals lives in this environment—animals that may carry one of two alleles, A_1 or A_2. Animals with the first allele eat (I_1) plants with smooth leaves and ignore (I_2) plants with wrinkled leaves, whereas animals with the second allele do the opposite. To keep things simple, it is assumed that the action of eating a plant reliably leads to the consequence of gaining the nutritional value of the plant, which is correlated with some fitness value for the organism in question. In effect, consequents are perfectly correlated with interpretants (and independent of objects).

Of interest is the evolution of this population, given this environment and the relative fitness of such alleles. We will use equation (4.1), along with equation (4.4), to answer the following questions: How does the frequency of alleles in the population change over time? Where are the fixed points? What is the nature of their stability?

Given the scenario just described, the evolution of the animal population turns on a set of parameterized relations. First, there is the relative abundance of plants of either type in the environment (which is assumed to be fixed on the time scales in question):

$$P(O) = \begin{bmatrix} p & 1-p \end{bmatrix}. \tag{4.7}$$

There is the likelihood that a plant of one type or the other will have smooth or wrinkled leaves:

$$P(S/O) = \begin{bmatrix} 1 & 0 \\ 0 & 1 \end{bmatrix}. \tag{4.8}$$

There is the probability of a consequent (gaining a plant's nutritional value or not), given an interpretant (eating a plant or ignoring it):

$$P(C/I \wedge O) = P(C/I) = \begin{bmatrix} 1 & 0 \\ 0 & 1 \end{bmatrix}. \tag{4.9}$$

There is the fitness value gained as a function of the type of plant eaten (where a_1 and a_2 are positive real numbers):

$$V(C \wedge I \wedge O) = V(C \wedge O) = \begin{bmatrix} a_1 & a_2 \\ 0 & 0 \end{bmatrix}. \tag{4.10}$$

There is the probability that an animal with the A_1 allele will eat or ignore a plant, depending on whether it has smooth or wrinkled leaves:

$$P_{A_1}(I/S) = \begin{bmatrix} 1 & 0 \\ 0 & 1 \end{bmatrix}. \tag{4.11}$$

Finally, there is the probability that an animal with the A_2 allele will eat or ignore a plant depending on whether it has smooth or wrinkled leaves:

$$P_{A_2}(I/S) = \begin{bmatrix} 0 & 1 \\ 1 & 0 \end{bmatrix}. \tag{4.12}$$

Having specified all the values in these matrices in accordance with our description of the scenario in question, equation (4.4) may be used to calculate the relative fitness of each allele. After doing some algebra, we find the following values:

$$F(A_1) = p \cdot a_1 \,, \tag{4.13a}$$

$$F(A_2) = (1-p) \cdot a_2 \,. \tag{4.13b}$$

These relative fitnesses, in conjunction with equation (4.1) and the normalization constraint $P(A_1) + P(A_2) = 1$, may now be used to calculate the changing frequencies of alleles in the population (given some initial set of frequencies). In light of our discussion at the end of section 4.1, we know that the population will have the following overarching evolutionary dynamics. First, if the parameter values are such that $F(A_1) = F(A_2)$, the frequencies of alleles will remain fixed at their initial values. If $F(A_1) > F(A_2)$, then the population will evolve toward $P(A_1) = 1$. Finally, if $F(A_1) < F(A_2)$, then the population will evolve towards $P(A_2) = 1$.

Setting the two fitnesses equal to each other and solving for p, we find the following threshold value for the relative frequency of the two plants (p versus $1-p$), given their nutritional values (a_1 and a_2) and the likelihood $P(S/O)$ that they have smooth or wrinkled leaves:

$$p_{crit} = \frac{a_2}{a_1 + a_2} \,. \tag{4.14}$$

If $p > p_{crit}$, animals bearing allele A_1 will take over the population. If $p < p_{crit}$, animals bearing allele A_2 will take over the population. Finally, if $p = p_{crit}$, both alleles will remain in the population at their initial frequencies. In other words, even if plants with wrinkled leaves are more nutritious than plants with smooth leaves, so long as plants with smooth leaves are abundant enough in the environment that $p = P(O_1) > p_{crit}$, then animals with the A_1 allele will take over the population, insofar as that allele generates semiotic processes that better tap into—and hence are better adapted to—the nutritional value and relative abundance of the relevant objects in that environment.

Figure 4.1 shows several trajectories for this animal population as a function of different plant abundances in the environment. The parameters have been chosen so that O_2 plants are slightly more nutritious than O_1 plants, and hence slightly more fitness-promoting ($a_2 > a_1$). As may be seen by the dotted line in the figure, even if the initial frequency of A_1 alleles is arbitrarily close to zero, so long as $p > p_{crit}$ organisms with this allele will eventually take over the population.

We have been assuming that the alleles cause their bearers to engage in all-or-nothing interpretants, and hence behaviors that are maximally contrastive. In the case of A_1, the bearer eats plants with smooth leaves and ignores plants with wrinkled leaves; in the case of A_2, the bearer eats plants with wrinkled

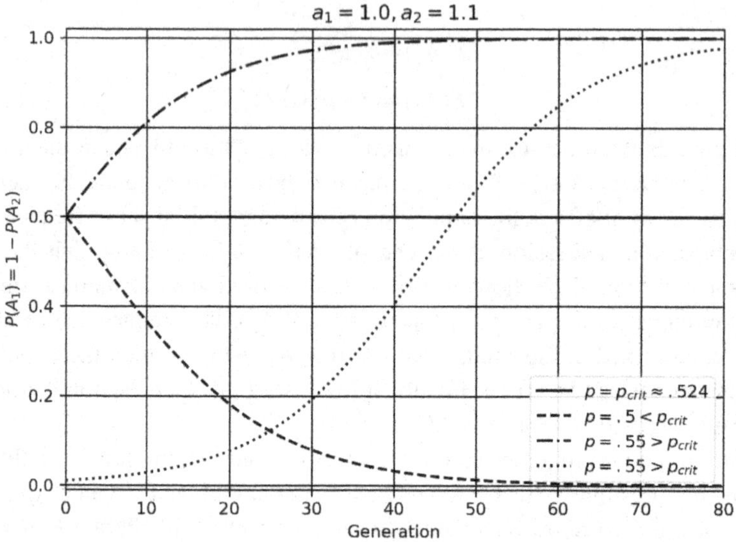

Figure 4.1: Frequency of Alleles over Generations

leaves and ignores plants with smooth leaves. It should be emphasized that even if agents bearing allele A_1 were only the slightest bit more likely to eat smooth-leaved plants than agents bearing allele A_2, the same fixed points and threshold values would be found. Suppose, for example, that agents bearing allele A_2 eat plants randomly, such that

$$P_{A_2}(I/S) = \begin{bmatrix} 1/2 & 1/2 \\ 1/2 & 1/2 \end{bmatrix}. \tag{4.15}$$

In contrast, suppose that agents bearing allele A_1 are attracted to plants with smooth leaves just a tiny bit more than they are attracted to plants with wrinkled leaves, such that

$$P_{A_1}(I/S) = \begin{bmatrix} 1/2+\epsilon & 1/2-\epsilon \\ 1/2-\epsilon & 1/2+\epsilon \end{bmatrix}, \tag{4.16}$$

where $0 < \epsilon \leq 1/2$.

In such a scenario, no matter how close to zero ϵ becomes, we will find the same fixed points and critical threshold as the scenario just analyzed, and hence nearly identical evolutionary dynamics. In other words, so long as $p > p_{crit} = a_2/(a_1 + a_2)$, bearers of the A_1 allele will take over the population completely. The only difference will be the speed, in terms of number of generations, that it takes for this to happen. The evolutionary dynamics of this scenario, as well as its underlying tendencies, are thus quite robust.

This example, as simple as it is, sheds light on one possible mode of biosemiotic meaning. Every organism within a population may be considered a semiotic agent in its own right, insofar as the allele that it carries causes it to cause particular interpretants in the context of particular signs (as triggered and triggering events, respectively), insofar as the objects causing these signs and the consequents caused by these interpretants relate in such a way that they contribute to the value obtained by the organism bearing the allele in question—and hence to the fitness of the allele per se. Insofar as organisms have been selected to bear alleles of one type rather than another, as a function of how the semiotic process caused by the allele affects the relative fitness of its bearer by causing them to become more frequent in the population over time, it seems unproblematic to call these "biosemiotic processes" and to call such emergent effects "biosemiotic adaptation." To be sure, the fact that the population had these two alleles in its gene pool in the first place was not necessarily selected for, but might rather have been the result of some chance mutation, founder effects, cataclysmic events, and other factors. This section has only examined movement towards local optima, given the alleles that were present in the population.

In this scenario, the plants themselves (as types of objects with relatively correlated types of signs and certain value-correlated properties) were not semiotic agents. They simply happened to embody a correlation that the interpreting animals could exploit. They thereby constituted a relatively fixed environment that shaped the distribution of alleles in the animal population. They were, in this last sense, causal agents (insofar as they, or at least their properties, changed the frequency of alleles in the animal population), but they were not biosemiotic agents (insofar as they were not themselves selected to do this).

In the next section, we will examine a scenario in which both plants and animals are biosemiotic agents (as well as causal agents), insofar as each population is selected by the other and selecting of the other.

4.4 Dynamic Organism, Dynamic Environment

We will now examine an extension of the preceding scenario in which not only are the plant-eating animals biosemiotic agents, but the plants themselves are as well. In particular, just as there are two alleles in the population of animals, A_1 and A_2, there are two alleles in the population of plants, O_1 and O_2. Whereas the former continue to cause their bearers to instigate certain interpretants in the context of certain signs, the latter simply cause their bearers to evince particular signs. In this way, the plants are signers and the animals are interpreters. These relations may be formalized as follows:

$$A_1 \rightarrow P_{A_1}(I/S)\,, \tag{4.17a}$$

$$A_2 \rightarrow P_{A_2}(I/S)\,, \tag{4.17b}$$

$$O_1 \rightarrow P_{O_1}(S)\,, \tag{4.17c}$$

$$O_2 \rightarrow P_{O_2}(S)\,. \tag{4.17d}$$

Note, then, that we continue to deal with organisms that have one relevant locus with two alleles. Now, however, both the plants and the animals are semiotic agents, insofar as their signifying and interpreting practices, or semiotic strategies, were mutually selected.

In keeping with the earlier scenario, suppose that plants with the O_1 allele always have smooth leaves (S_1) and plants with the O_2 allele always have wrinkled leaves (S_2), such that

$$P_{O_1}(S) = \begin{bmatrix} 1 & 0 \end{bmatrix}, \tag{4.18a}$$

$$P_{O_2}(S) = \begin{bmatrix} 0 & 1 \end{bmatrix}. \tag{4.18b}$$

Or, equivalently, both equations may be combined to get

$$P(S/O) = \begin{bmatrix} 1 & 0 \\ 0 & 1 \end{bmatrix}. \tag{4.19}$$

Suppose that animals with the A_1 allele eat (I_1) plants with smooth leaves and ignore (I_2) plants with wrinkled leaves, whereas animals with the A_2 allele do the opposite, such that

$$P_{A_1}(I/S) = \begin{bmatrix} 1 & 0 \\ 0 & 1 \end{bmatrix}, \tag{4.20a}$$

$$P_{A_2}(I/S) = \begin{bmatrix} 0 & 1 \\ 1 & 0 \end{bmatrix}. \tag{4.20b}$$

Suppose that eating a plant reliably correlates with acquiring the plant's nutrition (and damaging the plant), whereas ignoring a plant reliably correlates with not acquiring the plant's nutrition (and leaving the plant unscathed), such that

$$P(C/I \wedge O) = P(C/I) = \begin{bmatrix} 1 & 0 \\ 0 & 1 \end{bmatrix}. \tag{4.21}$$

Finally, whereas the fitness of the animals is increased by eating the plants, the fitness of the plants is decreased by being eaten by the animals, such that the change in relative fitness due to such interactions is

$$V_A(C \wedge I \wedge O) = V_A(C \wedge O) = \begin{bmatrix} a_1 & a_2 \\ 0 & 0 \end{bmatrix}, \tag{4.22a}$$

$$V_O(C \wedge I \wedge O) = V_O(C \wedge O) = \begin{bmatrix} -b_1 & -b_2 \\ 0 & 0 \end{bmatrix}. \tag{4.22b}$$

Having specified all the values in these matrices in accordance with our description of the scenario in question, equation (4.4) may be used, along with its coupled analog, to calculate the relative fitness of each allele in both populations (due to their mutual interactions):

$$F(A_n) = \sum_{ijkl} P(O_i) \cdot P(S_j/O_i) \cdot P_{A_n}(I_k/S_j) \cdot P(C_l/I_k) \cdot V_A(C_l \wedge O_i), \tag{4.23a}$$

$$F(O_n) = \sum_{ijkl} P(S_j/O_n) \cdot P(A_i) \cdot P_{A_i}(I_k/S_j) \cdot P(C_l/I_k) \cdot V_O(C_l \wedge O_n). \tag{4.23b}$$

As may be seen, the equation for $F(O_n)$ complements the equation for $F(A_n)$. If the latter specifies the fitness of an interpreting agent, the former specifies the fitness of a signifying agent, and as the latter depends on the frequency of the former via the term $P(O_i)$, the former depends on the frequency of the latter via the term $P(A_i)$. In this way, the fitness of alleles in each population depends on the frequency of alleles in the other population, such that the evolution of each population is dynamically coupled to the evolution of the other.

Keeping in mind that there are now two normalization constraints, $P(O_1) + P(O_2) = 1$ and $P(A_1) + P(A_2) = 1$, equations (4.23a) and (4.23b) may be used to calculate the relative fitness of each of the alleles, due to their participation in this division of semiotic labor. Undertaking the algebra yields

$$F(A_1) = a_1 \cdot P(O_1), \tag{4.24a}$$

$$F(A_2) = a_2 \cdot [1 - P(O_1)], \tag{4.24b}$$

$$F(O_1) = -b_1 \cdot P(A_1), \tag{4.24c}$$

$$F(O_2) = -b_2 \cdot [1 - P(A_1)]. \tag{4.24d}$$

Finally, it is assumed that something contributes to the fitness of members of both populations that is independent of, and/or prior to, their interaction with each other, and that is the same for both of the alleles in each population. For example, each population may have another source of nutrition that is relatively stable over the time scales in question. These will be considered their prior fitnesses: that is, the fitness of the organisms prior to their participation in such semiotic processes. This gives the final fitness equations for both populations:

$$F(A_1)=\alpha+a_1\cdot P(O_1)\,, \tag{4.25a}$$

$$F(A_2)=\alpha+a_2\cdot[1-P(O_1)]\,, \tag{4.25b}$$

$$F(O_1)=\beta-b_1\cdot P(A_1)\,, \tag{4.25c}$$

$$F(O_2)=\beta-b_2\cdot[1-P(A_1)]\,, \tag{4.25d}$$

where α and β, as the parameters representing the fitness values of the additional sources of nutrition, are assumed to be non-negative real numbers. We will typically assume that $b_1=b_2$, meaning that the fitness of plants with either allele are negatively affected by the same amount when eaten by animals. We will also assume that $\beta\geq b_1$ and $\beta\geq b_2$, ensuring that the relative fitness of plants, like the relative fitness of animals, is always greater than or equal to zero.

Insofar as there are now two populations of biosemiotic agents, there are two sets of coupled difference equations:

$$P'(A_i)=\frac{P(A_i)\cdot F(A_i)}{\sum_j P(A_j)\cdot F(A_j)}\,, \tag{4.26a}$$

$$P'(O_i)=\frac{P(O_i)\cdot F(O_i)}{\sum_j P(O_j)\cdot F(O_j)}\,. \tag{4.26b}$$

Setting $P'(A_i)=P(A_i)$ and $P'(O_i)=P(O_i)$, and undertaking some algebra, we find the following fixed points:

$$P(A_1)=0,\ P(O_1)=0\,; \tag{4.27a}$$

$$P(A_1)=0,\ P(O_1)=1\,; \tag{4.27b}$$

$$P(A_1)=1,\ P(O_1)=0\,; \tag{4.27c}$$

$$P(A_1)=1,\ P(O_1)=1\,. \tag{4.27d}$$

These points are, of course, relatively expected: they are simply those four points in phase space where each population has only alleles of one type or another. (Recall that $P(A_2)=1-P(A_1)$ and $P(O_2)=1-P(O_1)$ given the normalization constraints.) As will be seen, for the range of parameter values of interest, each of these four points attracts and repels trajectories in a counterclockwise pattern. For example, the fixed point given by $P(O_1)=0$ and $P(A_1)=0$ attracts trajectories on the line $P(O_1)=0$ and repels trajectories on the line $P(A_1)=0$. The fixed point given by $P(O_1)=1$ and $P(A_1)=0$ attracts trajectories on the line $P(A_1)=0$ and repels trajectories on the line $P(O_1)=0$. And it is similar for the other two fixed points.

There is one more fixed point with the following value:

$$P(A_1)=\frac{b_2}{b_1+b_2}\,,\quad P(O_1)=\frac{a_2}{a_1+a_2}\,. \tag{4.28}$$

As will be seen in the analysis that follows, while this fixed point is independent of the values of α and β, the behavior of each population around this point is sensitive to these values.

To gain insight into the behavior of this system, it is useful to study its dynamics for a range of extreme parameter values. We might begin by supposing that $\alpha \gg a_1, a_2$ and $\beta \gg b_1, b_2$. In such a situation, $F(A_1)\approx F(A_2)$, such that the frequency of each allele in the animal population stays constant. Similarly, $F(O_1)\approx F(O_2)$, such that the frequency of each allele in the plant population stays constant. This means that every possible point $P(O_1), P(A_1)$ in phase space is a fixed point, such that the system will remain exactly where it is, no matter where it begins.

Such a situation is relatively easy to understand: the fitness conferred upon the animals by eating the plants (as represented by a_1 and a_2) and the fitness taken from the plants by being eaten by the animals (as represented by b_1 and b_2) are negligible compared to the fitness that plants and animals get from other sources (as represented by α and β). Insofar as that fitness is the same for each allele in both populations (α for the interpreting animals, and β for the signifying plants), those populations undergo no salient evolutionary dynamics. While this limit is somewhat boring, it will serve as a useful baseline for understanding the behavior of the system for other parameter values.

Next, suppose that $\beta \gg b_1, b_2$. As before, $F(O_1)\approx F(O_2)$, such that the frequency of each allele in the plant population remains fixed at their initial values. No such condition is imposed upon the population of animals, however. Setting $F(A_1)=F(A_2)$, we find the following critical value:

$$P_{crit}(O_1)=\frac{a_2}{a_1+a_2}\,. \tag{4.29}$$

If $P_{in}(O_1)>P_{crit}(O_1)$, then $F(A_1)>F(A_2)$ for all values of $P(A_1)$ such that the A_1 allele will take over the population completely. Conversely, if $P_{in}(O_1)<$

$P_{crit}(O_1)$, then $F(A_1)<F(A_2)$ for all values of $P(A_1)$ such that the A_2 allele will take over the population completely. Finally, if $P_{in}(O_1)=P_{crit}(O_1)$, then $F(A_1)=F(A_2)$ for all values of $P(A_1)$ such that every value of $P(A_1)$ is a fixed point. In effect, the evolutionary dynamics are equivalent to those of the system discussed in the section 4.3.

Next, suppose that $\alpha \gg a_1, a_2$. Now $F(A_1) \approx F(A_2)$ such that the frequencies of all alleles in the animal population remain fixed at their initial values. No such condition is imposed upon the population of plants, however. Setting $F(A_1)=F(A_2)$, we find the following critical value:

$$P_{crit}(A_1) = \frac{b_2}{b_1 + b_2}. \tag{4.30}$$

If $P_{in}(A_1) > P_{crit}(A_1)$, then $F(O_1)<F(O_2)$ for all values of $P(O_1)$ such that the O_2 allele will take over the population completely. Conversely, if $P_{in}(A_1) < P_{crit}(A_1)$, then $F(O_1)>F(O_2)$ for all values of $P(O_1)$ such that the O_1 allele will take over the population completely. Finally, if $P_{in}(A_1)=P_{crit}(A_1)$, then $F(O_1)=F(O_2)$ for all values of $P(O_1)$ such that every value of $P(O_1)$ is a fixed point. In effect, the evolutionary dynamics are complementary to those of the system discussed in section 4.3: the plants, rather than the animals, are the biosemiotic agents; and the animals, rather than the plants, are the selecting environment.

Finally, suppose that $\alpha \geq a_1, a_2$ and $\beta \geq b_1, b_2$. Under these very general conditions, $P(A_1)$ oscillates around $P_{crit}(A_1)$ and $P(O_1)$ oscillates around $P_{crit}(O_1)$, where the amplitude of the oscillations increases over time. Figure 4.2 shows this behavior by graphing $P(O_1)$ and $P(A_1)$ over multiple generations, for a system with the following critical points: $P_{crit}(O_1)=.4$ and $P_{crit}(A_1)=.5$. Such oscillatory behavior makes sense, given how the scenario was first described: $P(O_1)$ increases when $P(A_1)$ is relatively small (because it is precisely animals with the A_1 allele that eat plants with the O_1 allele); however, this causes $P(A_1)$ to increase (as there is now a lot of food for the A_1-bearing animals to eat), which causes $P(O_1)$ to decrease (as there are now more animals eating O_1-bearing plants); and so it continues.

If we model this system (within a similar range of parameter values) as a differential equation rather than a difference equation, and take a linear approximation of the system for trajectories near the fixed point (as per the choice of initial conditions), we find that the period of oscillation (in units of generations) is

$$T = 2\pi\sqrt{\frac{(a_1\alpha + a_2\alpha + a_1 a_2)(b_1\beta + b_2\beta - b_1 b_2)}{a_1 a_2 b_1 b_2}}. \tag{4.31}$$

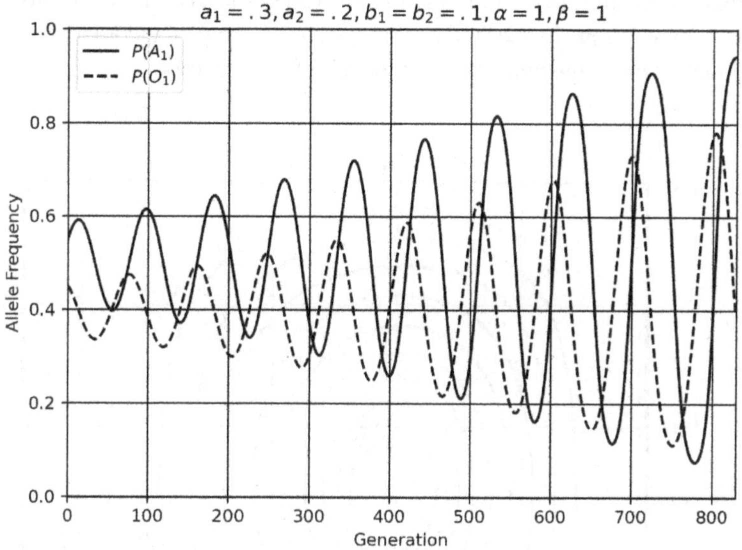

Figure 4.2: Evolution of System ($\alpha \geq a_1, a_2$ and $\beta \geq b_1, b_2$)

For the parameter values used in figure 4.2, for example, $T \approx 84$ generations, which is in pretty good agreement with the computationally generated results from the difference equation (in which there are about 10 periods in 840 generations). While equation (4.31) is somewhat unwieldy, it can easily be seen that, assuming $\alpha \approx \beta$, $\alpha \gg a_1, a_2$, and $\beta \gg b_1, b_2$, the period scales linearly with α (or β). In other words, there is a linearly increasing relation between the size of the external sources of fitness and the number of generations it takes for the system to return close to where it began. In contrast, when the external sources of fitness are small, then the coupled dynamics take over, and the system shifts from one extreme to the other on a much faster time scale.

We just plotted the frequency of alleles as a function of generation. As figure 4.3 shows, it is also illuminating to plot the *phase portrait* of the system, which shows the system moving through phase space over time. As may be seen, $P(O_1)$ is plotted along the x-axis and $P(A_1)$ is plotted along the y-axis. (Note that what might be initially understood as a four-dimensional phase space is really a two-dimensional phase space because of the two normalization constraints: $P(O_1) + P(O_2) = 1$ and $P(A_1) + P(A_2) = 1$.) The black line in the graph shows the movement of the system in phase space (which unfolds in a counterclockwise direction). And the vectors at each point in this space, given by $\Delta P(O_1) = P'(O_1) - P(O_1)$ and $\Delta P(A_1) = P'(A_1) - P(A_1)$, show the direction and relative magnitude of the system's movement at each point in this space across a generation. In particular, the system of coupled populations moves along paths whose tangent lines are given by these vectors.

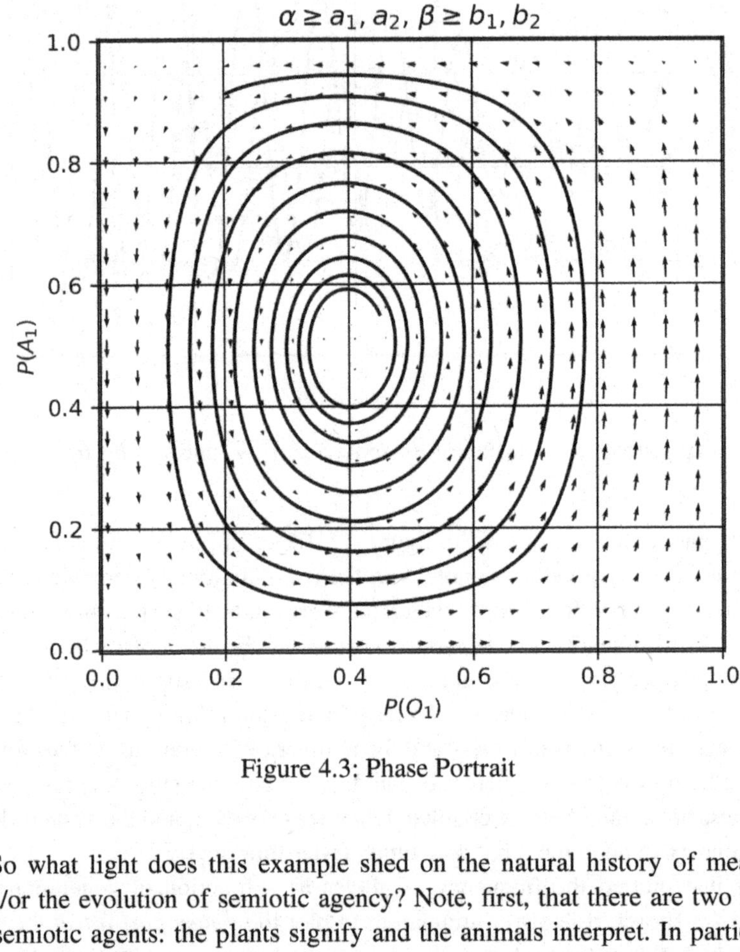

Figure 4.3: Phase Portrait

So what light does this example shed on the natural history of meaning and/or the evolution of semiotic agency? Note, first, that there are two kinds of semiotic agents: the plants signify and the animals interpret. In particular, each of these populations has two kinds of alleles. In the population of plants, alleles determine which kind of sign will be expressed. In the population of animals, alleles determine which of kind of interpretant will be triggered by such a sign. The relative frequency of each allele in a population depends on the relative fitness that it confers upon its bearer, which itself depends on the relative frequency of alleles in the other population. In effect, each population constitutes a constantly evolving environment that shapes the evolution of the other population, such that the system never really settles down.

It must be emphasized, then, that the individual plants and animals are not selecting which sign or interpretant to express: they simply exhibit the behavior/form that their allele determines. In this example, the object of semiosis was the plant allele in question, insofar as it was correlated with one leaf type or the other. In effect, the object of semiosis was the identity of the signifying agent. Interpreting agents that reacted to signs in ways that made them fitter (so far as their alleles disposed them to eat those plants that were more frequent and/or more nutritious) flourished—if only temporarily. And signifying agents that gave off signs that made them fitter (in the sense of being less likely to be eaten, given the relative frequency of each type of interpreting agent in the population) also flourished—again, if only temporarily. Note, then, how the signifying agents that did best were those that produced signs that made them more likely to be overlooked by the interpreting agents. In some sense, their strategy was akin to a mode of non-intentional deceit, unconscious camouflage, or teleonomic dissembling.

Having analyzed a case of competitive semiosis, we now investigate a case of cooperative semiosis: when coupled populations of signers and interpreters rise and fall together as a function of particular features of the languages that they employ, as well as the way that such features relate to the environment they inhabit.

4.5 Semiosis and Symbiosis

Suppose that there are two interacting populations, one consisting of signifying agents and the other consisting of interpreting agents. Members of both populations reside in an environment in which there is a certain fixed probability of finding food within a given generation. Whereas members of the first population can find the food (but not gather it), members of the second population can gather the food (but not find it). Organisms within the two populations need to communicate with each other to organize their cooperative efforts to obtain the value that the food provides.

The rest of this section will examine the evolutionary dynamics of these relatively symbiotic populations as a function of the cost and quality of the encoding and decoding schemes—or languages—employed by their members.

Suppose that there are two alleles in the signifying population (which we will call E_1 and E_2, for encoding alleles). Assuming that O_1 is the event of food being present and O_2 is the event of food being absent, and assuming that S_1 is the event of producing one sign (say, a dot, or some other behavior, sound, or signal) and S_2 is the event of producing another sign (say, a dash), the first two alleles cause their bearers to produce signs in the context of objects with the following likelihoods (understood as distinct encoding schemes):

$$E_1 \rightarrow P_{E_1}(S/O) = \begin{bmatrix} 1 & 0 \\ 0 & 1 \end{bmatrix}, \tag{4.32a}$$

$$E_2 \rightarrow P_{E_2}(S/O) = \begin{bmatrix} 0 & 1 \\ 1 & 0 \end{bmatrix}. \tag{4.32b}$$

In other words, signifying agents with allele E_1 indicate food with a dot (and its absence with a dash), whereas signifying agents with allele E_2 indicate food with a dash (and its absence with a dot). As will be discussed further shortly, we assume that dots are somewhat lighter than dashes: while they are cheaper to produce, they are more likely to be affected by external sources of noise.

Next, assume that there are two alleles in the interpreting population (which we will call D_1 and D_2, for decoding alleles). If I_1 is the event of coming to collect the food and I_2 is the event of staying put, these alleles cause their bearers to produce interpretants in the context of signs with the following likelihoods (understood as distinct decoding schemes):

$$D_1 \rightarrow P_{D_1}(I/S') = \begin{bmatrix} 1 & 0 \\ 0 & 1 \end{bmatrix}, \tag{4.33a}$$

$$D_2 \rightarrow P_{D_2}(I/S') = \begin{bmatrix} 0 & 1 \\ 1 & 0 \end{bmatrix}. \tag{4.33b}$$

In other words, interpreting agents with allele D_1 go to the signer if they sense a dot (and otherwise stay put), whereas interpreting agents with allele D_2 go to the signer if they sense a dash (and otherwise stay put).

We will assume that going to the signer (I_1) reliably correlates with arriving (C_1), such that interpreting agents that engage in this behavior not only expend energy by undertaking the journey, they are also well placed to help the signer obtain food should it exist. And we will assume that staying put (I_2) reliably correlates with going nowhere (C_2), such that interpreting agents who engage in this behavior expend no energy, but they also forgo any chance to help the signer acquire food. In effect, these are just simplifying assumptions, such that

$$P(C/I \wedge O) = P(C/I) = \begin{bmatrix} 1 & 0 \\ 0 & 1 \end{bmatrix}. \tag{4.34}$$

Insofar as there is a perfect correlation between interpretants and consequents, our specification of values—as given next—will only turn on consequents, signs, and objects.

Loosely, organisms with alleles E_1 and D_1 communicate by means of one language, whereas organisms with alleles E_2 and D_2 communicate by means of another language. Encoding signs with one scheme or the other is only of value if there are organisms that can decode the signs using a complementary scheme; and decoding signs with one scheme or the other is only of value if there are organisms who can encode the signs using a complementary scheme. Broadly, the fates of organisms carrying E_1 and D_1 alleles rise and fall together, as do the fates of organisms carrying E_2 and D_2 alleles. In what follows, we will refer to the complementary encoding and decoding schemes of E_1 and D_1 as "language 1" and to the complementary encoding and decoding schemes of E_2 and D_2 as "language 2."

As indicated by the difference in notation, it is assumed that the sign that the signifying agent sends (S) is not necessarily the same as the sign that the interpreting agent receives (S'). In particular, there may be noise in the channel that is employed by members of both populations. We will assume that the noise can affect each of the two signs differently, such that

$$P(S'/S) = \begin{bmatrix} 1-n_1 & n_2 \\ n_1 & 1-n_2 \end{bmatrix}. \tag{4.35}$$

When $n_1 = n_2 = 0$, there is no noise in the channel, such that received signs are identical to sent signs. As n_1 increases, what is sent as S_1 is more and more likely to be received as S_2. Similarly, as n_2 increases, what is sent as S_2 is more and more likely to be received as S_1. Finally, when $n_1 = n_2 = 1/2$, the channel is so noisy that there is no correlation between received and sent signs, and thus also no correlation between the signs that are received by the interpreter and the objects that they originally stood for.

It is assumed that each kind of sign has a cost that is borne by the signer: $c(S_1)$ and $c(S_2)$. As may be seen, these costs turn on the sign that is sent by the signifying agent, not the sign that is received by the interpreting agent. Similarly, each kind of interpretant has a cost that is borne by the interpreter: $c(I_1)$ and $c(I_2)$.

Aside from these role-specific costs, both agents (signer and interpreter) receive a value of $v/2$ if they successfully communicate, such that the interpreting agent comes to help (I_1) when food is found (O_1) by the signer. In other words, the fitness value of the food is v, and it is split evenly between the signer and interpreter when they successfully communicate about its presence, and thereby cooperatively coordinate their behavior to obtain it. It is assumed that the chance of finding food during a given generation is p, such that $P(O_1) = 1 - P(O_2) = p$.

Finally, it is assumed that both signifying agents have the same prior fitness (α), and both interpreting agents have the same prior fitness (β). These contributions to the overall fitness of such agents are independent of their participation in such communicative practices, insofar as they turn on other sources of food that such agents have access to. The larger α and β are, the less the evolutionary dynamics of such populations turn on their participation in such communicative practices.

Putting all these costs and benefits together, we find that the value that goes to signifying agents, as a function of the consequents of their behavior in relation to the behavior of interpreting agents in relation to the presence (O_1) and absence (O_2) of food, is as follows:

$$V_E(C \wedge S \wedge O_1) = \begin{bmatrix} \alpha + v/2 - c(S_1) & \alpha + v/2 - c(S_2) \\ \alpha - c(S_1) & \alpha - c(S_2) \end{bmatrix}, \tag{4.36a}$$

$$V_E(C \wedge S \wedge O_2) = \begin{bmatrix} \alpha - c(S_1) & \alpha - c(S_2) \\ \alpha - c(S_1) & \alpha - c(S_2) \end{bmatrix}. \tag{4.36b}$$

Similarly, and somewhat more simply, we find that the value that goes to the interpreting agents, as a function of the consequents of their behavior in relation to the presence and absence of food, is as follows:

$$V_D(C \wedge S \wedge O) = V_D(C \wedge O) = \begin{bmatrix} \beta + v/2 - c(I_1) & \beta - c(I_1) \\ \beta - c(I_2) & \beta - c(I_2) \end{bmatrix}. \tag{4.37}$$

In what follows, it is assumed that expressing the smaller sign (S_1) is cost-free, but the reliability of its transmission is easily affected by noise in the environment; in contrast, the larger sign (S_2) is costly to express, but the reliability of its transmission is unaffected by noise. For example, it may be that the larger sign has more momentum, and thus is less likely to be buffeted by wind, but it takes more energy to send. It is assumed that the action of staying put (I_2) is cost-free; in contrast, the action of coming to help (I_1) is costly. And while $c(S_2)$ and $c(I_1)$ are relatively costly, it is assumed that they are not more costly than the value obtained by signifying and interpreting agents when food is found and successfully gathered. In short, $c(S_1) = c(I_2) = 0$, whereas $0 \leq c(S_2) \leq v/2$ and $0 \leq c(I_1) \leq v/2$; and $n_2 = 0$, whereas $0 \leq n_1 \leq 1/2$.

Setting $\alpha = \beta \approx v$ ensures that the fitness conferred by cooperation through communication is maximally salient to the evolutionary dynamics of both populations. As will be seen, the values of these two variables do not change the position of the system's fixed points. However, their values speed up or slow down the rate of evolution, and hence the number of generations that it takes for the system to arrive at a fixed point.

Our final equations for the fixed points of the system will only depend on the values of v, $c(S_2)$, $c(I_1)$, n_1, and p. We are interested, then, in various trade-offs between the relative cost and noise sensitivity of the two languages in relation to the relative abundance of food in the environment (as the object of those languages), and how this affects the evolutionary dynamics of the populations themselves. In some sense, then, this scenario allows us to study the relation not just between meaning and motivation, but also between channels and codes. Even more broadly, we will be able to assess the relative utility, or viability, of different languages in different environments—and hence the degree to which one language or the other is more naturally suited to the features of the environment, given the needs of the agents.

Insofar as there are a lot of variables to keep track of, it is worthwhile summarizing them in a single table as follows:

Variable	Description	Value
p	Chance of finding food	$0 \leq p \leq 1$
v	Value of resource	$v = 1$
α	Prior fitness of signifying agents	$\alpha = v = 1$
β	Prior fitness of interpreting agents	$\beta = v = 1$
n_1	Probability that S_1 affected by noise	$0 \leq n_1 \leq 1/2$
n_2	Probability that S_2 affected by noise	$n_2 = 0$
$c(S_1)$	Cost of S_1 to signer	$c(S_1) = 0$
$c(S_2)$	Cost of S_2 to signer	$0 \leq c(S_2) \leq v/2$
$c(I_1)$	Cost of I_1 to interpreter	$0 \leq c(I_1) \leq v/2$
$c(I_2)$	Cost of I_2 to interpreter	$c(I_2) = 0$

The first column shows each of the key variables. The second column describes the dimension that each variable is meant to represent. Finally, the third column specifies the value, or range of values, that such variables can take in the scenario.

We will employ the usual equations to determine the frequency of alleles in the next generation of encoding and decoding agents, given the frequency and fitness of such alleles in the current generation:

$$P'(E_n) = \frac{P(E_n) \cdot F(E_n)}{\sum_m P(E_m) \cdot F(E_m)}, \tag{4.38a}$$

$$P'(D_n) = \frac{P(D_n) \cdot F(D_n)}{\sum_m P(D_m) \cdot F(D_m)}. \tag{4.38b}$$

Here are the equations that determine the fitness of alleles in the populations of encoding and decoding agents:

$$F(E_q) = \sum_{ijklmn} P(O_i) \cdot P_{E_q}(S_j/O_i) \cdot P(S'_k/S_j) \cdot P(D_l) \cdot P_{D_l}(I_m/S'_k) \cdot P(C_n/I_m) \cdot V_E(C_n \wedge S_j \wedge O_i), \tag{4.39a}$$

$$F(D_q) = \sum_{ijklmn} P(O_i) \cdot P(E_j) \cdot P_{E_j}(S_k/O_i) \cdot P(S'_l/S_k) \cdot P_{D_q}(I_m/S'_l) \cdot P(C_n/I_m) \cdot V_D(C_n \wedge O_i). \tag{4.39b}$$

While these two equations look somewhat complicated, they simply express the relative fitness of signifying and interpreting agents as a function of the environment in which they find themselves, an environment that turns on: the relative frequency of interpreters $P(D)$ and signers $P(E)$, the amount of noise in the channel $P(S'/S)$, the chance of finding a food source or not $P(O)$, and the likelihoods, eventualities, and values just discussed. They are simply well-motivated variations of our main equation as adapted to fitness calculations. Recall equation (4.4).

Pulling all the foregoing details together, we find five fixed points, where $P'(E_n) = P(E_n)$ and $P'(D_n) = P(D_n)$ for all n. The first four fixed points are the usual ones, where

$$P_f(E_1), P_f(D_1) = \begin{cases} (0,0), \\ (0,1), \\ (1,0), \\ (1,1). \end{cases} \tag{4.40}$$

In other words, and as expected, a signifying population that consists of all E_1 alleles, or all E_2 alleles stays that way (unless perturbed). And an interpreting population that consists of all D_1 alleles or all D_2 alleles, stays that way (unless perturbed). Crucially, assuming that the fifth fixed point, to be discussed next, is such that $0 < P_f(E_1) < 1$ and $0 < P_f(D_1) < 1$, the first and last of these are stable fixed points, whereas the middle two are unstable fixed points. In other words, almost all trajectories of the system move towards places where there are only E_1 and D_1 alleles or only E_2 and D_2 alleles; and then they remain there, and stably so. Such a result should not be too surprising: signifying and interpreting agents that use the same language to cooperatively solve a problem tend to be collectively successful, so evolutionary trajectories tend to move towards such positions in phase space.

The fifth and final fixed point is also unstable. It may be found by solving the coupled equations $F(E_1) = F(E_2)$ and $F(D_1) = F(D_2)$, under the usual normalization constraints: $P(E_1)+P(E_2)=1$ and $P(D_1)+P(D_2)=1$. While it is solvable, albeit messy, in the most general case (where none of the values of the variables have been specified), we will focus on a scenario in which $\alpha=\beta=v$, $c(I_1)=v/4$, and $v=1$. Undertaking the algebra, we find that

$$P_f(E_1)=\frac{1-2\cdot n_1+2\cdot n_1\cdot p}{2-2\cdot n_1}, \tag{4.41a}$$

$$P_f(D_1)=\frac{2\cdot c(S_2)-4\cdot c(S_2)\cdot p+p-n_1\cdot p}{2\cdot p-2\cdot n_1\cdot p}. \tag{4.41b}$$

In particular, $P_f(E_1)$ is the frequency of E_1 when $F(D_1)=F(D_2)$. Similarly, $P_f(D_1)$ is the frequency of D_1 when $F(E_1)=F(E_2)$. As will be seen, the position of this fifth, unstable fixed point, $(P_f(E_1), P_f(D_1))$, determines the relative attractiveness, or *viability*, of the two languages: What percentage of initial conditions flow to each of the two stable fixed points, such that language 1 completely takes over or language 2 completely takes over (corresponding to the two stable fixed points $(1, 1)$ and $(0, 0)$, respectively).

To understand the location of such fixed points and the nature of this system's behavior more generally, figure 4.4 plots a phase portrait of the system when $p=.7$, $n_1=.2$, and $c(S_2)=c(I_1)=1/4$. The thin solid lines, with arrows along them, indicate the general flow of trajectories starting at any point $(P(E_1), P(D_1))$ in phase space. As may be seen, almost all such trajectories move away from unstable fixed points, located at $(0, 1)$ and $(1, 0)$, and towards stable fixed points, located at $(0, 0)$ and $(1, 1)$. The four thick solid lines simply indicate actual trajectories, through phase space, of our two coupled populations, when starting with different initial conditions: $P(E_1)=.1$ and $P(D_1)=.1$, $P(E_1)=.1$ and $P(D_1)=.9$, $P(E_1)=.9$ and $P(D_1)=.1$, and $P(E_1)=.9$ and $P(D_1)=.9$. As may be seen, three of these trajectories move towards the stable fixed point $(1, 1)$, while the fourth trajectory moves towards the stable fixed point $(0, 0)$. Upon arriving, all four trajectories remain fixed in place.

As indicated by the intersection of the two dashed lines, besides the four fixed points at the corners, the fifth and final fixed point for these parameters is located at around $P_f(E_1)=.55$ and $P_f(D_1)=.32$. A system that starts there will stay there forever unless perturbed. Along the line $P(E_1)=P_f(E_1)$, indicated by the dashed vertical line, the trajectories are such that $P(D_1)=P(D_2)$. That is, all trajectories are horizontal. Along the line $P(D_1)=P_f(D_1)$, indicated by the dashed horizontal line, trajectories are such that $P(E_1)=P(E_2)$. That is, trajectories are vertical. Such horizontal and vertical lines, which are usually called "null-clines," reveal important aspects of a system's possible trajectories through phase space.

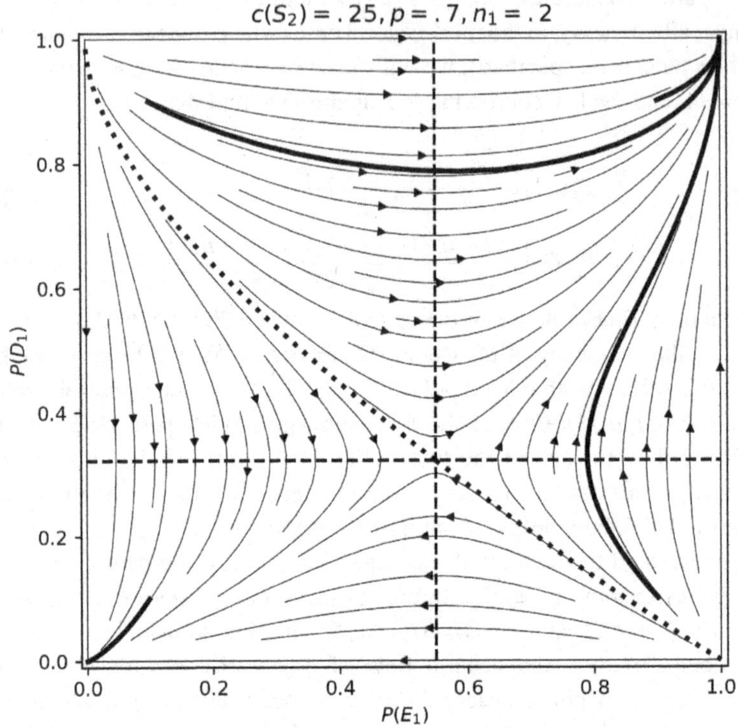

Figure 4.4: Key Features of Phase Portraits

As may be seen, these dashed lines carve phase space into four rectangles. All trajectories that start in, or pass through, the top-right rectangle go to $(1,1)$. All trajectories that start in, or pass through, the bottom-left rectangle go to $(0,0)$. In the top-left and bottom-right rectangles, it is harder to say, but about half of the trajectories in these two quadrants go to $(1,1)$, while the other half go to $(0,0)$.

Note that when the fifth fixed point is such that $P_f(E_1) < 0$ and/or $P_f(D_1) < 0$, the fixed point $(0,0)$ becomes unstable and almost all the trajectories end up at $(1,1)$. Conversely, when the fifth fixed point is such that $P_f(E_1) > 1$ and/or $P_f(D_1) > 1$, the fixed point $(1,1)$ becomes unstable and almost all the trajectories end up at $(0,0)$.

Very approximately, and assuming $0 \leq P_f(E_1) \leq 1$ and $0 \leq P_f(D_1) \leq 1$, such that the fifth fixed point in found within the realm of possible positions of the system, we can see that the basin of attraction for the point $(0,0)$ takes up a fraction of phase space that is about $\big(P_f(E_1) + P_f(D_1)\big)/2$, and hence the average of $P_f(E_1)$ and $P_f(D_1)$. Note, then, that as $P_f(E_1) \to 1$ and $P_f(D_1) \to 1$, the basin of attraction for $(0, 0)$ gets larger and larger (approaching 1). This means that the larger $P_f(E_1)$ and $P_f(D_1)$ are, the less attractive language 1 is (and the more attractive language 2 is), for the parameter values in question. Conversely, the smaller $P_f(E_1)$ and $P_f(D_1)$ are, the more attractive language 1 is (and the less attractive language 2 is). All this implies that if $P_f(E_1)$ goes up while $P_f(D_1)$ goes down, or vice versa, the basins of attraction for the two stable fixed points can remain relatively equal in size.

The dotted line indicates the actual (as opposed to approximate) separation between the basins of attraction. All the points in phase space that are above it end up at the stable fixed point $(1, 1)$, such that signers and interpreters using language 1 take over their respective populations completely (in particular, those carrying alleles E_1 or D_1). All the points in phase space that are below it end up at the stable fixed point $(0, 0)$, such that signers and interpreters using language 2 take over their respective populations completely (in particular, those carrying alleles E_2 and D_2). If we calculate the respective areas of the two basins of attraction, we see that about 39 percent of initial conditions end up at $(0, 0)$, and about 61 percent of initial conditions end up at $(1, 1)$. In other words, with this amount of noise, these costs for signs and interpretants, and this probability of finding food, signifying and interpreting agents that use language 1 win over their respective populations more often than signifying and interpreting agents that use language 2. As may also be seen, and as should be unsurprising, initial conditions with high frequencies of E_1 and D_1 tend to flow to $(1, 1)$; and initial conditions with low frequencies of E_1 and D_1, and thus high frequencies of E_2 and D_2, tend to flow to $(0, 0)$. In other words, the larger the community using a code, at the onset, the more likely that code will take over.

Note, finally, that—for these parameter values—there is no stable solution in which there is a mixing of codes, such that the signers use one language and the interpreters use another. Nor is there a solution, aside from the fifth unstable fixed point, in which both populations persist over generations with a mixture of alleles, such that users of two distinct languages—and hence members of two distinct language communities—stably coexist.

To give readers more examples of the behavior of such systems, figure 4.5 shows four additional phase portraits for various values of $c(S_2)$, n_1, and p. (As before, $c(I_1) = .25$ and $\alpha = \beta = 1$.) As we move from the phase portrait in figure 4.4 to the top row of phase portraits in figure 4.5, p becomes smaller. This means that food is less likely to be found, which means that agents with allele E_1 are more often using S_2, and agents with allele E_2 are more often using S_1. Insofar as S_2 is more costly than S_1, the language that more frequently uses it to encode messages (in this case, language 1) is less attractive, or viable. And, as may be seen, the attractiveness of language 1 goes from about .61 to .5 to .26 as p goes from .7 to .5 to .3. (And the ultimate destination of black trajectories, with the same initial conditions, follows suit.) Note, then, that language 1 is not better or worse than language 2 per se. When $c(S_2) = .25$ and $n_1 = .2$, language 1 is more attractive in an environment in which food is abundant, whereas language 2 is more attractive in an environment in which food is scarce.

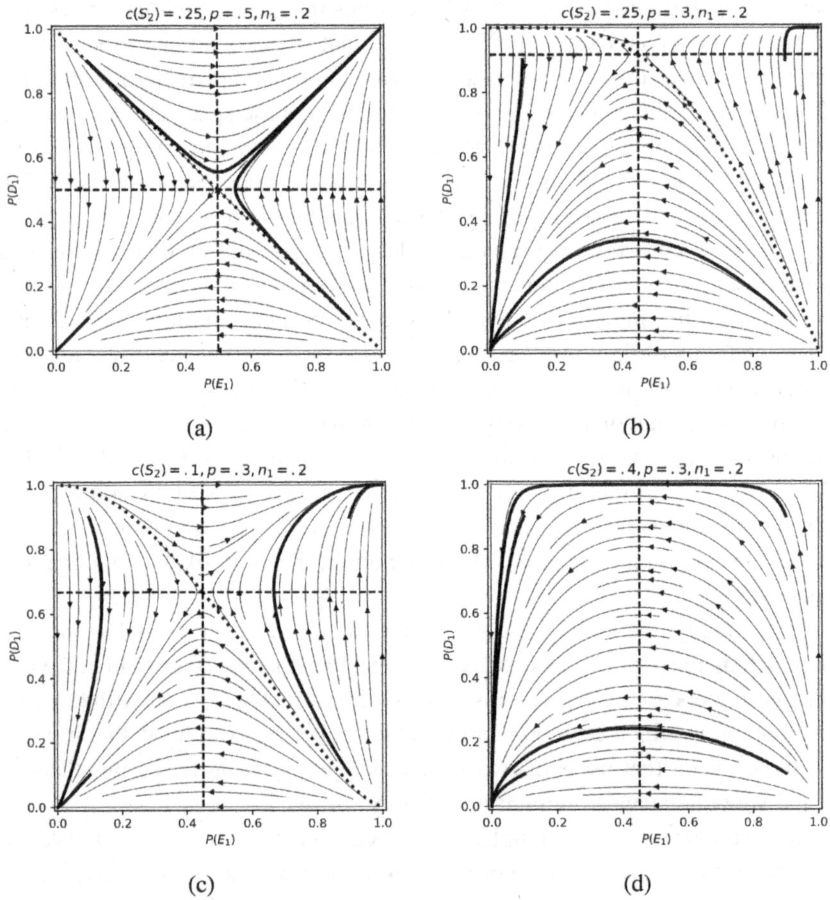

Figure 4.5: Four Phase Portraits Compared

In short, the language that better matches the environment—such that larger and more expensive signs are used infrequently and smaller and less expensive signs are used frequently, given the relative frequency of the objects being signified in that environment—is more attractive, in the sense that a larger percentage of initial positions in phase space have trajectories that eventually flow to it.

As we move from the top-right phase portrait of figure 4.5, to its bottom two phase portraits, we now keep the abundance of food and the amount of noise fixed ($p=3$ and $n_1=.2$), and we vary the cost of S_2. When $c(S_2)$ decreases (from .25 to .1), the attractiveness of language 1 increases (moving from .26 to .44). And when $c(S_2)$ increases (from .25 to .4), the attractiveness of language 1 decreases so much that almost no trajectories flow to $(1, 1)$. Indeed, in the bottom-right graph, the parameter values are such that $P_f(D_1)>1$, so there is no dashed horizontal line. Moreover, as may be seen by following the trajectory that starts at $(.9, .9)$ in the bottom-right phase portrait, even a population that starts with almost all E_1 and D_1 alleles ends up evolving towards all E_2 and D_2 alleles when the cost of S_2 is high enough (and the abundance of food is low, such that signers using language 1 are frequently encoding messages with the more expensive sign).

Rather than continue to plot individual phase portraits, it is more useful to plot the attractiveness of language 1 (itself equal to 1 minus the attractiveness of language 2), for a variety of parameter values. Recall that, with some caveats, the attractiveness of language 1 may be found by finding the area above the dotted lines in phase portraits such as these, insofar as that area holds all the points that eventually evolve towards the top-right corner (where language 1 takes over completely). The top and bottom graphs in figures 4.6 (a and b) show the attractiveness of language 1 as a function of the abundance of food (as parameterized by p) for various values of n_1 and $c(S_2)$, respectively.

Note, in particular, that the areas under these curves are a good measure of the overall attractiveness of language 1 across all environments (whereas the areas above the curves are a good measure of the overall attractivenes of language 2), and hence without reference to the particular details of any environment.

As may be seen in figure 4.6(a), language 1 is more attractive than language 2 in all environments in which food is abundant ($p>1/2$), and less attractive than language 2 in all environments in which food is scarce ($p<1/2$). As the sensitivity of S_1 to noise increases, the attractiveness of language 1 decreases for all values of p, but not by much. In part, this is because, even if the sign sent by agents bearing allele E_1 is disturbed by noise, agents bearing allele D_2 will still come to help, such that agents bearing allele E_1 end up obtaining the

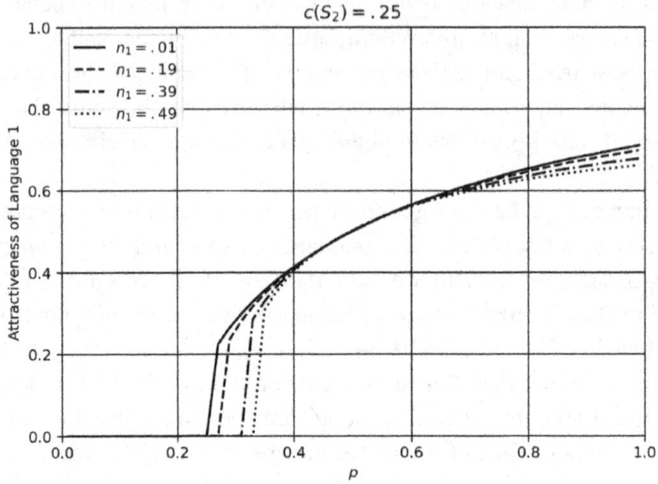

(a)

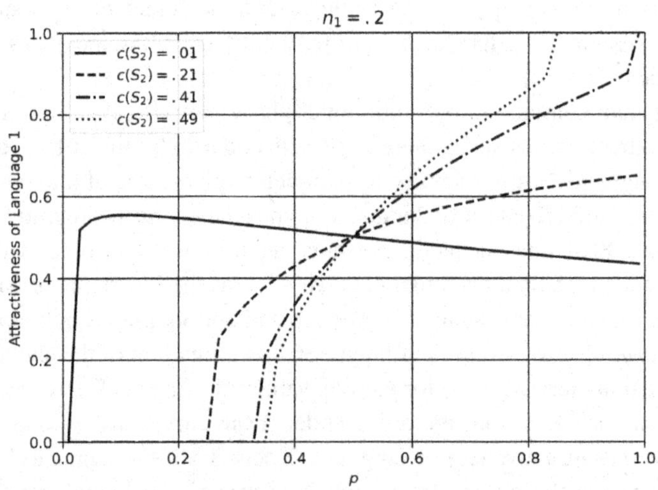

(b)

Figure 4.6: Attractiveness of Language 1

value provided by the food. Finally, if we look at the areas under these curves, which are always less than 1/2, we see that language 2 is more attractive than language 1 across all environments (at least for these conditions). Indeed, for small values of p, almost all possible initial conditions end up at $(0, 0)$, whereas for high p, only about 70 percent of initial conditions end up at $(1, 1)$.

As may be seen in figure 4.6(b), language 1 is again more attractive than language 2 in almost all environments in which food is abundant ($p > 1/2$), and less attractive than language 2 in all environments in which food is scarce ($p < 1/2$). As the cost of S_2 increases, the attractiveness of language 1 increases when $p > 1/2$ and decreases when $p < 1/2$. This should make sense, insofar as S_2 is expensive, the signers sending it suffer; and when $p > 1/2$, agents bearing allele E_2 tend to send it, whereas when $p < 1/2$, agents bearing allele E_1 tend to send it. This also makes the difference in cost much more salient at the extremes. As may be seen by the solid black line, if the cost of S_2 is small enough, then the noise-sensitivity of S_1 dominates, such that the agents using language 1 suffer when $p > 1/2$ and the agents using language 2 suffer when $p < 1/2$, insofar as—in such environments—those agents more frequently use the noisy sign. Finally, if we look at the areas under these curves, we again see that language 2 is generally more attractive than language 1 (under such conditions).

It is instructive to look at the phase portraits, and well-chosen sample trajectories, of systems near both sides of a kink in figure 4.6. Let us focus on two systems that would be found on the dotted line in figure 4.6(b), when $n_1 = 2$, $c(S_2) = .49$, and $p \approx .84$. Figure 4.7(a) shows the system when $p = .83$ and the fifth fixed point is located at $(.582, .013)$. Figure 4.7(b) shows the system when $p = .85$ and the fifth fixed point is located at $(.587, -.004)$. Also shown, in each phase portrait, is the trajectory of a system that starts at $(.05, .4)$. As may be seen, as soon as $P_f(D_1)$ drops below zero (when $p \approx .84$), the fixed point at $(0, 0)$ becomes unstable, such that systems that would have gone to $(0, 0)$ now go to $(1, 1)$. This also explains the fact that, near the end points in figure 4.6(b), the attractiveness of language 1 is no longer continuous: the system goes from an attractiveness of about .89 to an attractiveness of 1 as soon as $P_f(D_1)$ drops below zero. The small triangle of initial conditions that were going to $(0, 0)$ disappears all at once.

Note, then, that one language is not better than the other language per se. Language 1 serves its users better in an environment in which food is abundant, whereas language 2 serves its users better in an environment in which food is scarce (so long as S_2 is relatively expensive, and so long as S_1 is not too affected by noise). That said, across the range of parameter values that were examined, language 2 does better overall than language 1: its attractiveness, as averaged across all possible environments, was higher.

The next section contrasts the biosemiotic agents analyzed in this chapter with the rational agents analyzed in earlier chapters.

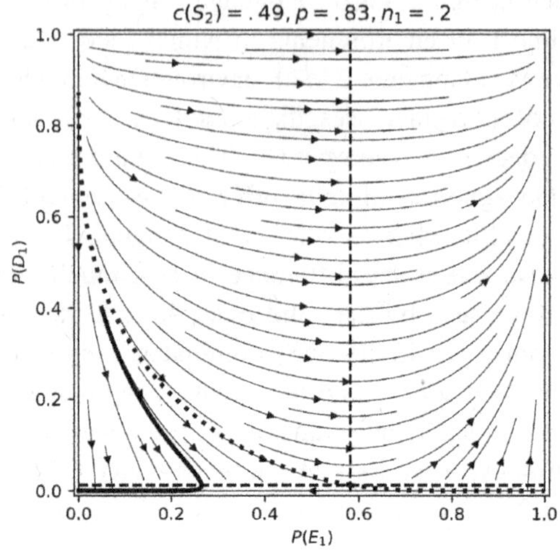

(a)

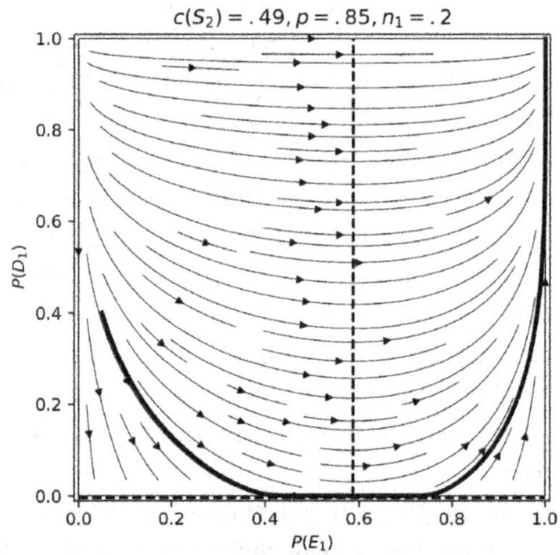

(b)

Figure 4.7: Behavior near Kinks

4.6 Meaning in Biosemiotic Systems

We have been focused on alleles that cause the organisms that carry them to cause particular interpretants in the context of particular signs (or cause particular signs in the context of particular objects). We stressed that such organisms do not need to perceive and act per se. All that matters is that they reliably trigger certain types of events (the interpretants) in the context of other types of events (the signs), where such signs are themselves triggered by other events (the objects) and such interpretants may be triggering of still other events (the consequents). So long as the relations between such triggering and triggered events, understood as the differential realization of particular worlds, contribute to the relative fitness of the allele in question, such that it is more likely to be present in the population in the next generation, such alleles may be called *biosemiotic agents.*

Unlike the relatively rational agents analyzed in chapters 2 and 3, such agents cannot be said to select particular interpretants in the context of particular signs (given their commitments and values). However, they may certainly be said to be selected by an environment, insofar as the patterns and propensities that they generate are in sync with, or fitted to, the causal patterns and valuable resources afforded by that environment. In effect, organisms bearing such alleles behave in ways that appear relatively rational, calculative, or ends-driven, even though the main causal mechanism underlying their emergence is sieving and serendipity (in the context of differential reproduction and inheritance).

If earlier chapters of this book focused on relatively subjective interpretive grounds, understood as the latent and labile principles guiding the behavior of more or less rational agents, this chapter has focused on relatively objective interpretive grounds. Rather than agent-specific sets of commitments and values, these consisted of environment-specific sets of relations between causal patterns and valuable resources. Insofar as the semiotic strategies—qua sign-interpretant relations—of organisms bearing certain alleles better tapped into such patterns and resources, such organisms were more likely to flourish, thereby contributing their alleles to the next generation.

While we focused on alleles that lead to sign-interpretant relations, we could have focused on alleles that altered any aspect of an organism's ability to engage in a fitness-altering semiotic process. Such alleles might change the size of an agent's ontology (i.e., what kinds of signs it can sense, and hence what kinds of objects it may relate to, as well as what kinds of interpretants it can instigate, and thus what kinds of consequents it can effect). They might change an organism's ability to use particular object-consequent relations (and thereby better secure valuable resources and/or tap causal patterns). They might make an organism more likely to send a signal that will be interpreted or to interpret a signal that has been sent. They might make an organism more likely to seek a mate that has a certain trait, or to be sought as a mate because of their traits, and so forth.

In many of the cases that were analyzed, the salient environments of biosemiotic agents consisted of the behavior of other biosemiotic agents. Concomitantly, the fitness of any agent's behavior, qua sign-interpretant patterning, turned on the relative frequency of other agents engaged in coupled behaviors with complementary patterns. In effect, semiotic agents selected, and were selected in turn, by other semiotic agents. Such couplings made the evolutionary trajectories of biosemiotic agents particularly complex, and often seemingly contradictory. For example, what counted as a useful—that is to say, fitness-inducing—behavior four generations ago could also cause, at some degree of remove, the removal of a key part of the environment that contributed to the fitness so induced. In this way, the apparent telos of such evolutionary trajectories was not just hard-earned, but sometimes short-lived.

We will continue to explore the evolution of biosemiotic agents in the next chapter, in which we examine the strategies of agents with different semiotic capacities, as they attempt to detect the identity of the agents around them (hawk or dove), to better cooperate—or compete—with them.

5 Hawks, Doves, and Mutants

Overview

This chapter relates the biosemiotic framework introduced in the previous chapter to key ideas in evolutionary game theory. It models the evolutionary dynamics of a population that has three kinds of agents (or alleles): hawks, doves, and mutants. Whereas the first two agents are identical to their counterparts in the classic scenario analyzed by the mathematical biologist John Maynard Smith, the last agent is a biosemiotic agent that can adjust its behavior depending on the appearance of its competitor. Section 5.1 reviews the dynamics of the classic scenario in this tradition: a non-semiotic competition between hawks and doves. Section 5.2 introduces a third agent into this scenario—a hawklike mutant with an adjustable semiotic capacity. Section 5.3 analyzes the fixed points of such a scenario and measures the robustness of mutants across a range of environments. Section 5.4 analyzes an analogous scenario, in which the mutant is dovelike rather than hawklike. And the final two sections study the fixed points and evolutionary dynamics of hawklike and dovelike mutants in scenarios where the semiotic ability of the mutant is more or less costly.

5.1 The Classic Scenario

Let us begin by reviewing the classic scenario introduced by Smith and Price (1973), and further developed by Smith (1982). Suppose that there is a population of interbreeding agents. Every generation, they engage in pairwise competitions over some sharable resource. In such competitions, one subtype in the population plays the relatively aggressive strategy of hawk, and the other subtype plays the relatively meek strategy of dove. (And aside from these differences in behavior, the two subtypes are identical in appearance.) While both hawks and doves initially display aggression in such competitions, hawks subsequently escalate their aggression, whereas doves back down in the face of such escalations.

Depending on the pairing of subtypes in a competition (e.g., hawk versus hawk, hawk versus dove, dove versus hawk, or dove versus dove), the agents playing the strategies receive more or less of the resource and are more or less likely to be hurt in the encounter. This affects their fitness, which affects their frequency in the next generation, which affects their fitness in that generation. And so it continues. We are interested in the evolutionary dynamics of such populations: the frequency and fitness of the subtypes over time (or across generations) as functions of the values being vied for.

The lowercase letters h and d will refer to the two types of behaviors, as well as to the underlying alleles that lead to such behaviors. The expressions $P_g(h)$ and $P_g(d)$ will refer to the frequencies of such alleles, or the probabilities of such behaviors, in the population. And the expressions $F_g(h)$ and $F_g(d)$ will refer to the relative fitness of the two alleles. These frequencies and fitnesses are subscripted with the letter g because they depend on the generation in question. Insofar as there are no other alleles in the population, $P_g(h) + P_g(d) = 1$ for all generations.

When a hawk encounters a dove in a pairwise competition, which occurs with frequency $P_g(h) \cdot P_g(d)$, the dove ultimately retreats, so the hawk acquires the entire resource, which has a value of v (understood as its contribution to the hawk's fitness). When a dove encounters a dove, which occurs with frequency $P_g(d) \cdot P_g(d)$, it ends up sharing the resource and thus acquires a value of $v/2$. When a dove encounters a hawk, which occurs with frequency $P_g(d) \cdot P_g(h)$, it ultimately flees and thus acquires a value of 0. And when a hawk encounters a hawk, which occurs with frequency $P_g(h) \cdot P_g(h)$, it ends up fighting with the other hawk, acquiring a value of $(v-c)/2$ on average, where c is the cost of injury. Such costs and benefits are summarized in the following payoff matrix, which shows the value that goes to the agent (or allele) on the left when confronting the agent at the top:

$V(A_1, A_2)$	h	d
h	$(v-c)/2$	v
d	0	$v/2$

The fitness of alleles in a given generation depends not just on the frequency of alleles in that generation, but also on the entries in the payoff matrix:

$$F_g(h) = V(h,h) \cdot P_g(h) + V(h,d) \cdot P_g(d) + F\,, \tag{5.1a}$$

$$F_g(d) = V(d,h) \cdot P_g(h) + V(d,d) \cdot P_g(d) + F\,. \tag{5.1b}$$

In other words, the relative fitness of hawks and doves is explicitly dependent on the relative frequency of hawks and doves (during a given generation), insofar as this determines how likely they are to encounter another hawk or dove in a pairwise competition, and hence how often they get one entry in the payoff matrix rather than another. As may be seen, it is assumed that both alleles have an additional source of fitness, F, which comes from other resources in the environment that both kinds of agents have equal access to (and which is itself generation independent). As discussed in chapter 4, such a value does not affect the fixed points of the scenario, but the time-to-equilibrium gets larger as it increases.

As per the discussion of evolutionary dynamics and the fitness equation in chapter 4, we know that we can find the frequency of hawks and doves in the next generation ($g' = g + 1$), as a function of their frequency and fitness in the current generation (g), using the following coupled difference equations:

$$P_{g'}(h) = \frac{P_g(h) \cdot F(h)}{\langle F_g \rangle}, \tag{5.2a}$$

$$P_{g'}(d) = \frac{P_g(d) \cdot F(d)}{\langle F_g \rangle}. \tag{5.2b}$$

The term in the denominator of these equations denotes the average fitness of the two alleles in the current generation:

$$\langle F_g \rangle = P_g(h) \cdot F(h) + P_g(d) \cdot F(d). \tag{5.3}$$

Using the convention that allele A_1 corresponds with h and allele A_2 corresponds with d, we can rewrite equations (5.1), (5.2), and (5.3) more compactly:

$$P_{g'}(A_i) = \frac{P_g(A_i) \cdot F(A_i)}{\sum_j P_g(A_j) \cdot F(A_j)}, \tag{5.4a}$$

$$F_g(A_i) = F + \sum_j V(A_i, A_j) \cdot P_g(A_j). \tag{5.4b}$$

Given some initial frequency of alleles (when $g = 0$, so to speak) and the normalization condition, $\sum_i P_g(A_i) = 1$, these equations are enough to determine the changing frequency of alleles across generations, and hence to track the evolutionary dynamics of the population. This relatively compact notation, while somewhat cumbersome for such a simple scenario, will prove useful in later sections of this chapter when we will need to handle larger numbers of alleles and more complicated kinds of behavior.

Following the discussion of evolutionary dynamics in chapter 4, we know that fixed points in phase space occur whenever the frequency of every allele in the next generation is equal to its frequency in the current generation, such that $P_{g'}(A_i) = P_g(A_i)$ for all alleles. Undertaking the algebra, we find that this condition occurs whenever the following equation holds:

$$P_g(h) \cdot \big(1 - P_g(h)\big) \cdot F_g(h) = P_g(h) \cdot \big(1 - P_g(h)\big) \cdot F_g(d)\,. \tag{5.5}$$

And this equation holds whenever any of the following three conditions is met: $P_g(h) = 1$ and $P_g(d) = 0$; $P_g(h) = 0$ and $P_g(d) = 1$; or $F_g(h) = F_g(d)$.

The first two conditions are simple enough to understand: a population that consists of only hawks, or only doves, stays that way. The third condition is a little more complicated. It says that so long as the fitness of the alleles is equal, their frequency will not change from whatever it currently is. Setting $F_g(h) = F_g(d)$, using the fact that $P_g(d) = 1 - P_g(h)$, and undertaking some more algebra, we find that this last condition is met whenever

$$P_*(h) = 1 - P_*(d) = \frac{v}{c}\,. \tag{5.6}$$

As may be seen, we will usually denote fixed points with a subscript star ($*$), thereby highlighting the fact that they are generation-independent positions in the phase space of the system.

More generally, the population exhibits the following behavior. If $v \geq c$, then hawks take over the population completely (so long as their initial frequency is greater than 0). In effect, the value of the resource is large enough to offset the cost of fighting, so the hawk strategy beats the dove strategy for all initial conditions. And if $v < c$, a population of hawks and doves will evolve until the frequency of hawks in the population is equal to the value of the resource divided by the value of the cost. In effect, hawks and doves can coexist, and stably so, so long as the proportion of hawks in the population is equal to the ratio of benefits to costs. We will usually take $c = 1$ in what follows and study the behavior of the population in the regime where $0 \leq v \leq 1$ (such that it is relatively costly to be a hawk).

Figure 5.1 shows various examples of such behavior. The six lines plot the frequency of hawks in the population, as a function of generation, for different initial frequencies and values of the resource. In particular, thin lines show trajectories in which $v = 3/10$ and thick lines show trajectories in which $v = 8/10$. Solid lines show trajectories in which the hawk allele had an initial frequency in the population of 9/10; dashed lines show trajectories in which the initial frequency of hawks was 5/10; and dotted lines show trajectories in which the initial frequency was 1/10. As may be seen, regardless of where they start, all trajectories of hawks move towards the fixed point 8/10 when $v = 8/10$; and all trajectories of hawks move towards the fixed point 2/10 when $v = 2/10$. (Recall that the frequency of doves is simply 1 minus the frequency of hawks; and note that we have set $c = 1$ and $F = 1$.)

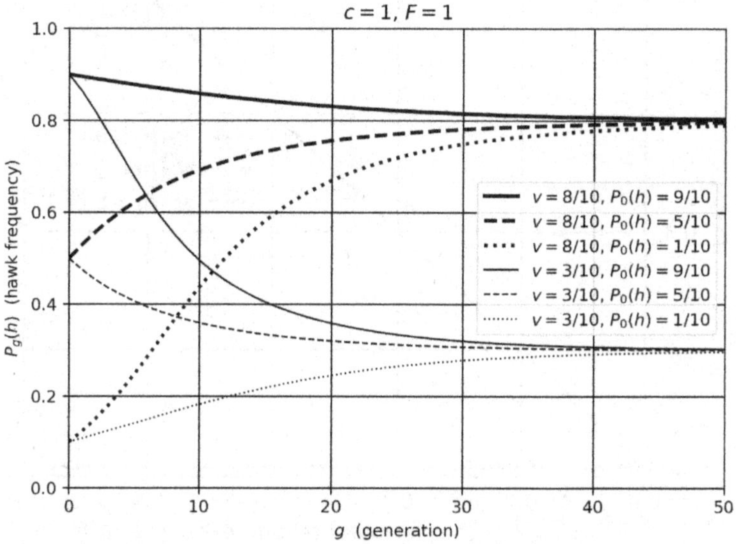

Figure 5.1: Temporal Dynamics of the Classic Hawk-Dove Scenario

Rather than plot actual trajectories, it is more useful to plot the fixed points of the system for they show the global behavior (qua threshold values and/or limit points) of the system, as a function of parameter values, regardless of the initial conditions. As discussed previously, and as seen in figure 5.2, this behavior is relatively simple. As shown by the dashed line, populations without doves stay that way (and hence remain fixed). As shown by the dotted line, populations without hawks stay that way. As shown by the solid line, populations with a mixture of hawks and doves evolve until the proportion of hawks in the population is v/c, and then they stay that way. Note that only the fixed points of hawks $P_*(h)$ are shown; the fixed points of doves are, of course, simply $1 - P_*(h)$.

The foregoing results are well known, following the classic work of Smith and Price. We plot these figures and go through this analysis because the classic scenario will show up as a limited case of the more complicated and inherently biosemiotic scenario that will be discussed next.

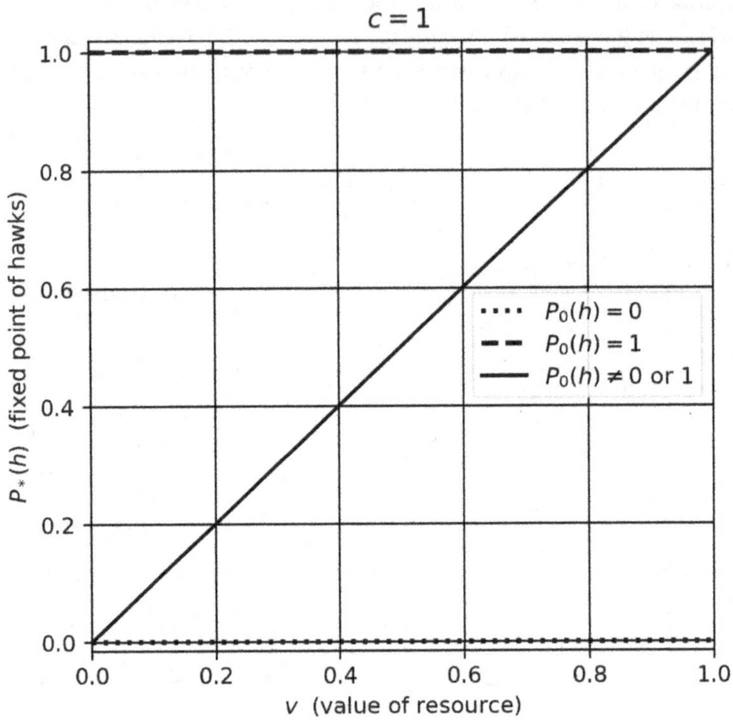

Figure 5.2: Fixed Points of the Classic Hawk-Dove Scenario

5.2 Biosemiotic Mutations

We are now going to introduce a third agent into the scenario, and hence a third allele into the population: the mutant (m). This is a biosemiotic agent that has the ability (however slight) to differentially relate to the specific identity of its competitor, and thereby alter its own behavior. In effect, it can semiotically exploit the patterns that are found in its environment—however unconsciously, uncalculatingly, and/or innately.

We assume, in particular, that hawks and doves continue to behave exactly as they did in the classic scenario. While hawks and doves are quasi-identical to each other in appearance, there is a perceivable trait that hawks are more likely to exhibit than doves, such that the presence or absence of the trait provides evidence of its bearer's identity. While hawks and doves cannot take this trait into account (either because they cannot perceive it or because they cannot alter their behavior), mutants behave slightly differently depending on whether their competitor does or does not exhibit the trait. It should be stressed that they do not do this intentionally or as the result of some unconscious calculation; it is simply that the allele that they carry causes them to behave differently. Whether such a mutant will flourish or founder in the population, and how much so, insofar as such behavior makes it more or less fit than the other alleles, is the question to be answered.

More specifically, mutants may be understood as slight mutations of hawks. They have the exact same appearance as hawks (and hence exhibit the sensible trait with the same frequency), and almost the same behavior. That is, they will usually threaten and then escalate when competing for the resource, just as hawks do. However, insofar as they also have the ability to sense the trait of their opponent, and adjust their behavior accordingly, they do not always behave like hawks. In particular, while they continue to behave as a hawk if their opponent does not exhibit the trait (and so appears to be a dove), they have a slight tendency to behave like a dove if their opponent exhibits the trait (and so appears to be a hawk).

It will prove convenient to denote the three alleles (and/or agents) as follows: $A_1 = h, A_2 = d, A_3 = m$. Using this convention, the relative frequencies (or probabilities) of such alleles, in a given generation g, will be denoted by $P_g(A_i)$, and their relative fitnesses will be denoted by $F_g(A_i)$. The average fitness of alleles in the population during a given generation is thus

$$\langle F_g \rangle = \sum_i F_g(A_i) \cdot P_g(A_i) , \tag{5.7}$$

where the frequencies of the alleles are normalized such that

$$\sum_i P_g(A_i) = P_g(A_1) + P_g(A_2) + P_g(A_3) = 1 . \tag{5.8}$$

We will continue to assume that we can calculate the relative frequency of such alleles in the next generation, as a function of their frequency and fitness in the current generation, using the following coupled difference equations:

$$P_{g'}(h) = \frac{F_g(h) \cdot P_g(h)}{\langle F_g \rangle} , \tag{5.9a}$$

$$P_{g'}(d) = \frac{F_g(d) \cdot P_g(d)}{\langle F_g \rangle} , \tag{5.9b}$$

$$P_{g'}(m) = \frac{F_g(m) \cdot P_g(m)}{\langle F_g \rangle} . \tag{5.9c}$$

Or, putting these ideas together, and letting the indices i and j range from 1 to 3, we have the following somewhat compact formulation:

$$P_{g'}(A_i) = \frac{F_g(A_i) \cdot P_g(A_i)}{\sum_j F_g(A_j) \cdot P_g(A_j)} . \tag{5.10}$$

Before we calculate the fitness of an allele, we need to specify the effect that each allele has on the organism that carries it. As discussed previously, but now more formally, it is assumed that the alleles slightly alter both the behavior and appearance of the agents that carry them (which are otherwise identical). This biosemiotic mediation can be expressed as follows:

$$A_i \rightarrow \begin{cases} P_{A_i}(S) , \\ P_{A_i}(I/S) . \end{cases} \tag{5.11}$$

In other words, allele A_i causes its bearer to evince certain signs (that other agents can sense); and it causes its bearer to instigate certain behaviors (when sensing certain signs).

As for appearance, agents carrying the three alleles—now understood as objects of semiotic processes—evince a certain trait with different frequencies. For the present purposes, exactly what the trait consists of does not matter so much: it might mean that the agent is slightly larger, or has a particular crest or a distinctive eye color. As such, it is not so much the trait per se, but rather its presence (S_1) or absence (S_2), that constitutes the signs in question. Organisms bearing the hawk allele are likely to exhibit the trait. Organisms bearing the dove allele are unlikely to exhibit the trait. Finally, organisms bearing the mutant allele, insofar as they are hawklike, are likely to exhibit the trait. These likelihoods may be expressed as follows (where $1/2 < a_1 \leq 1$, $a_2 = 1 - a_1$, and $a_3 = a_1$):

$$P_h(S) = \begin{bmatrix} a_1 \\ 1 - a_1 \end{bmatrix} , \tag{5.12a}$$

$$P_d(S) = \begin{bmatrix} a_2 \\ 1 - a_2 \end{bmatrix} = \begin{bmatrix} 1 - a_1 \\ a_1 \end{bmatrix} , \tag{5.12b}$$

$$P_m(S) = \begin{bmatrix} a_3 \\ 1 - a_3 \end{bmatrix} = \begin{bmatrix} a_1 \\ 1 - a_1 \end{bmatrix} . \tag{5.12c}$$

Having taken care of appearance, let us now turn to behavior. It is assumed that there are two kinds of strategic roles, or interpretants, that an agent can instigate: behave like a hawk (I_1) or behave like a dove (I_2). Organisms with the h allele only ever behave like hawks (regardless of whether their competitor exhibits the trait). Organisms with the d allele only ever behave like doves (again, regardless of the other organism's appearance). Finally, organisms with the m allele behave like hawks in the context of agents that do not have the trait (and thereby look like doves), and have a slight tendency to behave like doves in the context of agents that have the trait (and thereby look like hawks). These semiotic strategies may be expressed as follows (where $0 \leq e_1 \leq 1$ and $e_2 = 1$):

$$P_h(I/S) = \begin{bmatrix} 1 & 1 \\ 0 & 0 \end{bmatrix}, \tag{5.13a}$$

$$P_d(I/S) = \begin{bmatrix} 0 & 0 \\ 1 & 1 \end{bmatrix}, \tag{5.13b}$$

$$P_m(I/S) = \begin{bmatrix} e_1 & e_2 \\ 1-e_1 & 1-e_2 \end{bmatrix} = \begin{bmatrix} e_1 & 1 \\ 1-e_1 & 0 \end{bmatrix}. \tag{5.13c}$$

While the appearance, $P_{A_i}(S)$, and behavior, $P_{A_i}(I/S)$, of the agents bearing the alleles are described using five variables $(a_1, a_2, a_3, e_1, e_2)$, only the value of two of these variables, a_1 and e_1, can be independently specified in the current scenario. This is because hawks contrast with doves and mutants are hawklike. We will later analyze a scenario in which the mutant is dovelike. And other scenarios are also imaginable. All five variables are included at this early stage to make salient contrasts more evident and the analysis more general.

Following the classic hawk versus dove scenario from evolutionary biology and game theory, as discussed in section 5.1, we will assume the usual payoff matrix:

$$V(I_i, I_j) = \begin{bmatrix} \frac{v-c}{2} & v \\ 0 & \frac{v}{2} \end{bmatrix}. \tag{5.14}$$

Each entry in this matrix shows the value (qua fitness) that goes to an agent that is instigating I_i in the context of an agent that is instigating I_j. In particular, when an agent that behaves like a hawk encounters another agent that behaves like a hawk, it receives an average payout of $(v-c)/2$, where v is the value of the resource and c is the cost of the fighting (due to injuries sustained or energy expended). When an agent that behaves like a hawk encounters an agent that behaves like a dove, the payout is v, reflecting the fact that it receives the entire resource. When an agent that behaves like a dove encounters an agent that behaves like a hawk, it receives a value of 0, reflecting the fact that the other agent receives the entire resource. Finally, when an agent that behaves like a dove encounters another agent that behaves like a dove, the average payout is $v/2$, reflecting the fact that the two agents share the resource equally.

Note, then, that the payoff to a hawk or dove when it confronts a mutant will depend on the behavior of the mutant, which depends on whether the hawk or dove exhibits the trait. Similarly, the payoff to a mutant depends on whether it behaves like a hawk or a dove, which depends on whether its competitor (which may be another mutant) exhibits the trait.

Insofar as there are a lot of variables and cases to keep track of, it is worthwhile summarizing them in a single table, as follows:

Variable	Description	Value
c	Cost of fighting	$c=1$ (all scenarios)
v	Value of resource	$0\leq v\leq 1$ (all scenarios)
F	Fitness of alleles from other sources	$F=1$ (all scenarios)
k	Cost of mutant's semiotic ability	$0\leq k\leq 1$ (all scenarios)
a_1	Probability hawk evinces trait	$a_1>1/2$ (all scenarios)
a_2	Probability dove evinces trait	$a_2<1/2$ (all scenarios)
a_3	Probability mutant evinces trait	$a_3=a_1$ (hlm), $a_3=a_2$ (dlm)
e_1	Probability mutant acts like hawk if trait	$e_1<1$ (hlm), $e_1=0$ (dlm)
e_2	Probability mutant acts like dove if trait	$e_2=1$ (hlm), $e_2>0$ (dlm)

The first column shows each of the key variables. The second column describes the dimension that each variable is meant to represent. The third column specifies the value, or range of values, that such variables can take in the scenario in question. The expression *hlm* denotes hawklike mutants (as will be analyzed in section 5.3); and the expression *dlm* denotes dovelike mutants (as will be analyzed in section 5.4).

Taking into account such parameter-specific dependencies, we may now use the foregoing equations, in conjunction with a variant of our main equation (from chapter 1), to specify the fitness of each of the three alleles as follows:

$$F_g(A_i)=F+\sum_{jklmno} P_g(A_j)\cdot P_{A_j}(S_k)\cdot P_{A_i}(I_l/S_k)\cdot P_{A_i}(S_m)\cdot P_{A_j}(I_n/S_m)\cdot V(I_l,I_n)\,. \tag{5.15}$$

In other words, the fitness of allele A_i in a given generation g is equal to a summation over the product of the following six terms: the frequency of other alleles (and hence agents) in the population $P_g(A_j)$; the probability that such agents evince a certain sign $P_{A_j}(S_k)$; the probability that the agent in question instigates a certain interpretant in the context of such signs $P_{A_i}(I_l/S_k)$; the probability that the agent bearing the allele in question evinces a certain sign $P_{A_i}(S_m)$; the probability that the other agents instigate a certain interpretant in the context of such a sign $P_{A_j}(I_n/S_m)$; and the value that goes to an agent instigating I_l in the context of an agent instigating I_n. The indices used to sum over agents (objects or alleles) go from 1 to 3. The indices used to sum over signs and interpretants go from 1 to 2. In effect, then, we are simply summing over the values that go to each agent as a function of the kinds of interactions, semiotic or otherwise, that it has with other agents. As before, term F represents the contribution to the fitness of each allele that comes from other resources that all three agents have equal access to. As before, its value will not change the fixed points of the scenario (in the sense of the equilibrium states that the system tends to); however, as it increases, the time-to-equilibrium also increases, meaning that the number of generations that it takes to reach a fixed point gets greater and greater.

Before analyzing the evolutionary trajectories and fixed points of such a population, we need to fill in some more details about the scenario in question. Rather than study the fixed points of such a scenario for all possible parameter values, we are going to focus on a salient subset of the possibility space. We set $c=1$ and assume that $0 \leq v \leq 1$, such that the value of the resource is less than or equal to the cost of fighting. We initially assume that there are no mutants, such that the population of hawks and doves has come to equilibrium. Now, even though hawks and doves are different in appearance (in contrast to the classic scenario), that difference does not make them interact any differently as both such agents behave rigidly. So, without mutants, this scenario reduces to the classic scenario, such that the final (and fixed) fraction of hawks in the population is equal to the ratio of benefits to costs: $P_*(h) = v/c = v$.

We now assume that the third allele, itself a mutant of the hawk allele (insofar as its bearers are identical to hawks in appearance and almost identical to hawks in behavior), is introduced into this otherwise stable population. We assume, in particular, that it is introduced with some tiny, almost negligible frequency (as a chance mutation from some hawk forebearer), and the initial frequencies of the other two alleles are almost exactly as they would otherwise be (had they come to equilibrium, at the classic fixed points, for the value v in question). In particular, we assume that $P_0(h) = v - \epsilon$, $P_0(d) = 1 - v$, and $P_0(m) = \epsilon$, where ϵ is some arbitrarily small, positive quantity.

We want to know how such a mutant allele will do in such a context. Can it survive at all, or does it get sieved out of the population entirely? Are there any values of v when it can stabilize in the population or even take over the population completely? Can it push out, in the sense of extinguish, either of the other two alleles? And how do such facts change as we change the appearance of the agents (by changing the values of a_i), or change the behavior of mutants (by adjusting e_i)?

The next section will attempt to answer these questions.

5.3 Hawklike Mutants

The evolutionary dynamics of this three-agent scenario are relatively complicated, insofar as they are sensitive to the values of the parameters (v, c, F, a_i, e_i), as well as the initial frequencies of the alleles. To get some insight into the space of possible evolutionary trajectories, we will focus on fixed points: places in phase space where $P_{g'}(A_i) = P_g(A_i)$ for each of the three alleles, such that the population stops evolving. Like the classic case, but generalized, this should occur when any one of the following seven conditions occur:

1. $P_g(d) = P_g(m) = 0$ and $P_g(h) = 1$
2. $P_g(h) = P_g(m) = 0$ and $P_g(d) = 1$
3. $P_g(h) = P_g(d) = 0$ and $P_g(m) = 1$
4. $P_g(m) = 0$, $P_g(h) \neq 0$ and $P_g(d) \neq 0$, and $F_g(h) = F_g(d)$
5. $P_g(d) = 0$, $P_g(h) \neq 0$ and $P_g(m) \neq 0$, and $F_g(h) = F_g(m)$
6. $P_g(h) = 0$, $P_g(d) \neq 0$ and $P_g(m) \neq 0$, and $F_g(d) = F_g(m)$
7. $P_g(h) \neq 0$, $P_g(d) \neq 0$, $P_g(m) \neq 0$ and $F_g(h) = F_g(d) = F_g(m)$

The first three conditions are easy to understand: a population that consists of only hawks (or only doves, or only mutants) stays that way. The second three sets of conditions are somewhat easy to understand. If a population has no mutants and the frequencies of hawks and doves are non-zero, then it will have a fixed point where the fitness of hawks is equal to the fitness of doves. This is equivalent to the classic scenario. If a population has no doves and the frequencies of hawks and mutants are non-zero, then it will have fixed points where the fitness of hawks is equal to the fitness of mutants. And if a population has no hawks and the frequencies of doves and mutants are non-zero, then it will have fixed points where the fitness of doves is equal to the fitness of mutants. Finally, if a population has non-zero frequencies of hawks, doves, and mutants, its fixed points will occur when the fitnesses of hawks, doves, and mutants are simultaneously equal.

It should be emphasized that there is no guarantee that the fitness of one allele will be equal to the fitness of any other allele within the range of acceptable allele frequencies. And hence there is no guarantee that there will be any or all such fixed points (aside from the trivial ones). Finally, we have not said anything yet about the stability of such fixed points. Suffice it to say, most of the ones analyzed here will turn out to be stable.

As discussed in section 5.2, we want to study the space of possible trajectories when $c = 1$ and $0 \leq v \leq 1$ (such that it is somewhat costly to be a hawk), when $a_1 = a_3 > 1/2$ and $a_2 = 1 - a_1 < 1/2$ (such that S_1 correlates with hawks and mutants and S_2 correlates with doves), and when $e_2 = 1$ (such that the mutant not only looks like a hawk but also behaves in a relatively hawklike fashion). As in the classic case, the value of F does not affect the location of the critical points, only the speed (in generational time) with which the system arrives at them. In effect, then, there are only three salient variables with adjustable values: v, a_1, and e_1.

The first three conditions are relatively trivial, so nothing more will be said about them in this discussion. The last condition turns out to yield no possible solutions (where the values of $P_*(h)$, $P_*(d)$, and $P_*(m)$ are all greater than zero and less than 1) for the range of parameter values in question. In effect, there is no fixed point, stable or unstable, where all three alleles can coexist. The second three conditions are quite interesting, so we will focus on their elucidation in the rest of this section.

The first of these last three conditions, when $P_0(m)=0$ and $F_g(h)=F_g(d)$, is equivalent to the classic scenario (as should be expected). As shown in the last section, assuming $c=1$, so long as $0 \leq v \leq 1$, there is a stable fixed point when $P_*(d)=1-P_*(h)$ and

$$P_*(h)=v\,. \tag{5.16}$$

Plotted as a function of v, this is simply the equation of a line whose slope is 1 and whose y-intercept is zero (recall the solid line in figure 5.2).

The second of these last three conditions, when $P_0(d)=0$ and $F_g(h)=F_g(m)$, occurs when $P_*(h)=1-P_*(m)$ and

$$P_*(m)=\frac{1-v}{a_1\cdot(1-e_1)}\,. \tag{5.17}$$

Plotted as a function of v, this is also the equation of a line—one whose x-intercept is 1 and whose slope is a function of a_1 and e_1 (and is always negative for the range of parameter values in question). For points on this line where $P_*(m)$ is between 0 and 1 (and $0<v<1$), the hawk and mutant share phase space. For other points within this range of v values, the mutant takes over the population completely. In effect, the smaller the value of v, the better a hawklike mutant does against a hawk.

The third of these last three conditions, when $P_0(h)=0$ and $F_g(d)=F_g(m)$, occurs when $P_*(d)=1-P_*(m)$ and

$$P_*(m)=\frac{v\cdot(a_1+e_1-a_1\cdot e_1)}{a_1^2\cdot e_1^2-2\cdot a_1^2\cdot e_1+a_1^2+2\cdot a_1\cdot e_1-2\cdot a_1+1}\,. \tag{5.18}$$

Plotted as a function of v, this too is the equation of a line—one whose y-intercept is zero and whose slope depends on the values of a_1 and e_1 (and is always positive for the range of parameter values in question). For points on this line where $P_*(m)$ is between 0 and 1 (and $0<v<1$), the dove and mutant share phase space. For other points within this range of v values, the mutant takes over the population completely. In effect, the larger the value of v, the better a hawklike mutant does against a dove.

Crucially, neither of these last two functions describes, by itself, the fixed points of the scenario in question. Taken together, however, they precisely describe the fixed points of the three-allele scenario. To see how, look at figure 5.3, which plots these last two fixed points, both of which are stable, as a function of the value of resource v for various values of a_1 and e_1. (Recall that $c=1$, $e_2=1$, $a_2=1-a_1$, and $a_3=a_1$.)

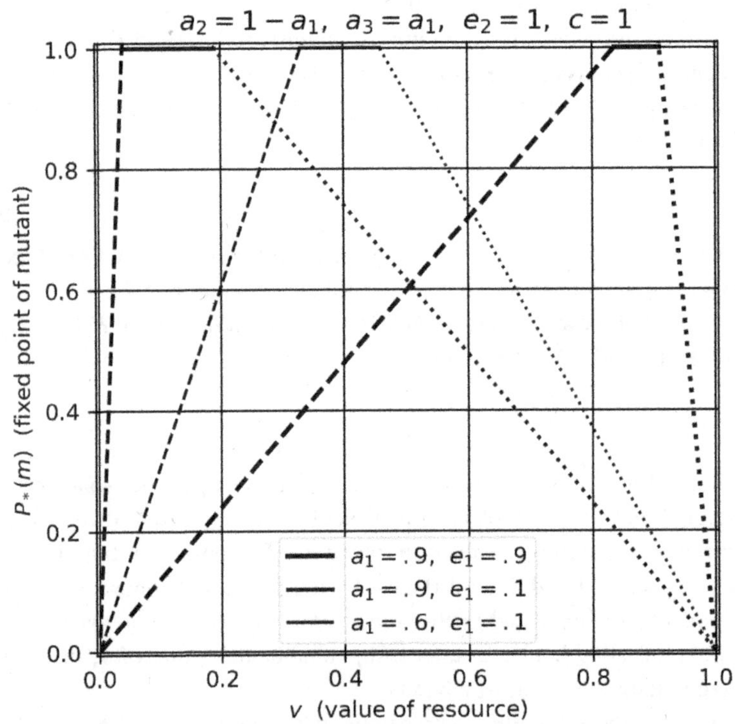

Figure 5.3: Fixed Points of Hawklike Mutants

To understand this somewhat complicated diagram, focus first on the thickest line, which represents the fixed points of the mutant when $a_1=.9$ (meaning that there is a strong correlation between S_1 and hawks as well as mutants, as well as between S_2 and doves), and $e_1=.9$, meaning that the mutant usually behaves like a hawk, even in the context of competitors that look like hawks. The dashed part of this line (which has a positive slope) shows the region where the mutant and dove share phase space (and the hawk allele has been sieved from the population). The fixed point of the mutant, as a function of v, is shown by the height of the dashed line as given by equation (5.18); and the fixed point of the dove is, of course, 1 minus that value. The dotted part of this line (which has a negative slope) shows the region where the mutant and hawk share phase space (and the dove allele has been sieved from the population). The fixed point of the mutant, as a function of v, is shown by the height of the dotted line as given by equation (5.17); and the fixed point of the hawk is, of course, 1 minus that value. Finally, the solid part of this line (which has zero slope) shows the region where the mutant takes over the population completely (such that both hawks and doves go extinct). Similar claims hold for the other two lines, showing how the size and slope of the three regions of fixed points shift as a function of the parameter values.

As may be seen, there are three key domains of behavior: a leftmost region, in which the value of the resource is low compared to the cost of fighting and the hawk allele cannot exist; a rightmost region, in which the value of the resource is high compared to the cost of fighting and the dove allele cannot exist; and a middle region, in which the value of the resource is such that the neither the hawk nor the dove can exist and the mutant takes over completely. Note how doves beat not only hawks, but also mutants (in the sense that they take over most of the population), when v is small; and hawks beat not only doves, but also mutants, when v is large. Recall, finally, that the mutants start off with near-zero frequency in the population, so the mutant allele is extremely viable—able to survive, or take over, in all three regions.

A useful measure of what might be called the *robustness* of an allele, $R(A_i)$, is its average final frequency in the population, and hence its fixed-point value when averaged across all values of v between zero and 1. Loosely, robustness is a measure of an allele's ability to flourish across a wide range of environments. In the classic scenario, in which there was only a hawk and a dove, each had a robustness of 1/2 (the area under or over, respectively, the solid line in figure 5.2). In figure 5.3, the robustness of the mutant, $R(m)$, is equivalent to the area under any one of the trapezoids delineated by the mutant's fixed points (for various values of the parameters a_1 and e_1). As may be seen, in the three-allele scenario, the robustness of the mutant is greater than 1/2 (for all the parameter values shown), whereas the robustness of both the hawk and dove combined is 1 minus this value, and thus less than 1/2. Not only does the mutant now do better than either the hawk or dove did in the two-allele scenario, it also does better than both of them combined in the three-allele scenario.

We have just explored one little portion of a much larger space: the part where mutants are hawklike in appearance ($a_3 = a_1$), hawklike in behavior when encountering an agent who is dovelike in appearance ($e_2 = 1$), and somewhat likely to be dovelike in behavior when confronting an agent that is hawk-like in appearance ($e_1 < 1$). In effect, it was assumed that mutants were a mutation of hawks, and hence otherwise very similar to them.

We now turn to mutants that are dovelike in appearance and behavior. As will be seen, while a dovelike mutant is not as robust as a hawk-like mutant, there are portions of parameter space where all three alleles (hawk, dove, and dovelike mutant) can stably coexist.

5.4 Dovelike Mutants

Let us now examine a scenario in which mutants are dovelike in appearance ($a_3 = a_2 = 1 - a_1$), dovelike in behavior when encountering an agent that is hawklike in appearance ($e_1 = 0$), and somewhat likely to be hawklike in behavior when confronting an agent that is dovelike in appearance ($e_2 > 0$). As in the case of hawklike mutants, it is assumed that the mutant is introduced into the population with an arbitrarily small initial frequency, and the initial frequencies of hawks and doves are essentially those that they would have in the classic scenario for particular values of v. We are interested in finding the fixed points of the population, as a function of v, for various values of the parameters a_1 and e_2.

The general conditions for fixed points are the same as they were in the preceding scenario, so we need not review them here. Suffice it to say that we are particularly interested in the last three conditions, so fixed points involving mixtures of hawks and dovelike mutants, mixtures of doves and dovelike mutants, and mixtures of hawks, doves, and dovelike mutants.

Let us first find the fixed points of mutants and doves when there are no hawks. Setting $P_g(h) = 0$ and solving for $P_g(m)$ when $F_g(m) = F_g(d)$, we find the following fixed point:

$$P_*(m) = 1 - P_*(d) = \frac{v}{a_1 \cdot e_2}. \tag{5.19}$$

As may be seen, this is just the equation for a line passing through the origin. When $a_1 = 1$ and $e_2 = 1$, such that neither mutant nor dove exhibits the trait and the mutant always acts like a hawk, the mutant does the same as a hawk against a dove, so we get something resembling the classic scenario. As a_1 decreases and/or e_2 decreases, such that mutants and doves sometimes exhibit the trait and/or the mutant sometimes behaves like a dove when confronting an agent that is dovelike in appearance, then the line of fixed points gets steeper. In effect, a dovelike mutant does even better against a dove than a classic hawk for small values of v and takes over the population completely when $v > a_1 \cdot e_1$.

Next, we find the fixed points of mutants and hawks when there are no doves. Setting $P_g(d) = 0$ and solving for $P_g(m)$ when $F_g(m) = F_g(h)$, we find the following fixed point:

$$P_*(m) = 1 - P_*(h) = \frac{-(a_1 \cdot e_2 - e_2 + 1) \cdot v + a_1 \cdot e_2 - e_2 + 1}{a_1^2 \cdot e_2^2 + 2 \cdot a_1 \cdot e_2 - 2 \cdot e_2 + 1}. \tag{5.20}$$

While this equation looks somewhat complicated, it too is just the equation of a line—one that always has a negative slope and passes through the point $(1, 0)$. It is relevant for our scenario because, for larger values of v, the mutant and the hawk can make the dove go extinct. It will be discussed next.

Finally, and perhaps most interestingly, we find that there are solutions where $F_g(m) = F_g(h) = F_g(d)$, such all three alleles can stably coexist. Solving for the fixed-point frequency of each of the three alleles when this condition is met, we find

$$P_*(h) = \frac{a_1 \cdot v \cdot (v+1)}{2 \cdot a_1 \cdot v^2 - v^2 + 1}, \tag{5.21a}$$

$$P_*(d) = \frac{a_1 \cdot e_2 \cdot v^2 - a_1 \cdot e_2 \cdot v - e_2 \cdot v^2 + e_2 + v^2 - v}{e_2 \cdot (2 \cdot a_1 \cdot v^2 - v^2 + 1)}, \tag{5.21b}$$

$$P_*(m) = \frac{v \cdot (1 - v)}{e_2 \cdot (2 \cdot a_1 \cdot v^2 - v^2 + 1)}. \tag{5.21c}$$

Using equations (5.21a, b, and c), figure 5.4 plots the fixed points of dovelike mutants (solid), hawks (dashed), and doves (dotted) as a function of v for various values of e_2 and a_1. In figures 5.4(c) and 5.4(d), where e_2 is relatively large, all three alleles can stably coexist for all values of v. In figures 5.4(a) and 5.4(b), where e_2 is relatively small, in contrast, there are two general regions: a leftmost region (where v is small), in which hawks, doves, and mutants coexist; and a rightmost region (where v is large), in which the dove has been sieved from the population, such that equation (5.20) takes over and only hawks and dovelike mutants remain. As may be seen, the transition happens when the dotted line (corresponding to the fixed point of doves) drops to zero, which corresponds to the following value of v:

$$v_{crit} = \frac{e_2}{a_1 \cdot e_2 - e_2 + 1}. \tag{5.22}$$

As may be seen in figure 5.4(a), figure 5.4(b), and figure 5.4(c), increasing e_2 lowers the robustness of mutants. And as may be seen in figure 5.4(c) and figure 5.4(d), decreasing a_1 raises the robustness of doves and dovelike mutants

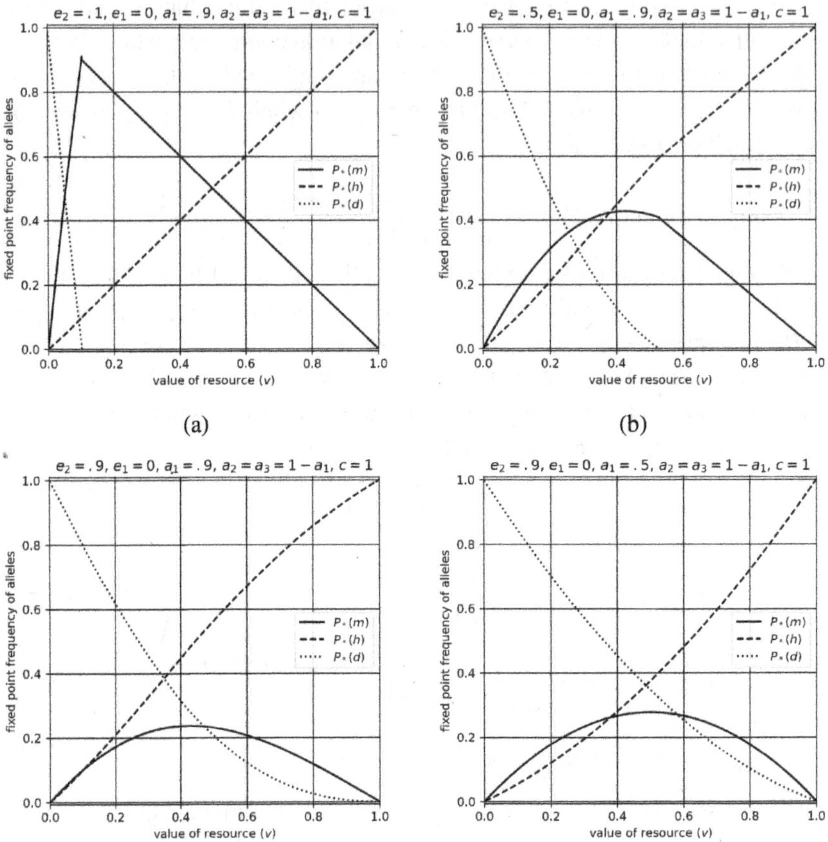

(a) (b)

(c) (d)

Figure 5.4: Fixed Points of Dovelike Mutants

and lowers the robustness of hawks. In short, unlike the last scenario, in which hawklike mutants were more robust than hawks and doves together, dovelike mutants never take over the population entirely.

After reviewing the results so far, the next section introduces a scenario in which it is relatively costly to have a semiotic capacity.

5.5 The Cost of a Semiotic Capacity

Readers may have noticed by now that there are really four kinds of mutants that can be examined with this framework:

1. Those that look exactly like hawks ($a_3 = a_1$), and typically behave like hawks, but have a tendency to behave like doves when confronting an organism that looks like a hawk ($e_1 < 1$, $e_2 = 1$)
2. Those that look exactly like hawks ($a_3 = a_1$), and typically behave like hawks, but have a tendency to behave like doves when confronting an organism that looks like a dove ($e_1 = 1$, $e_2 < 1$)
3. Those that look exactly like doves ($a_3 = a_2$), and typically behave like doves, but have a tendency to behave like hawks when confronting an organism that looks like a hawk ($e_1 > 0, e_2 = 0$)
4. Those that look exactly like doves ($a_3 = a_2$), and typically behave like doves, but have a tendency to behave like hawks when confronting an organism that looks like a dove ($e_1 = 0, e_2 > 0$)

The first case was examined in section 5.3, and the fourth case was examined in section 5.4. The reason why we did not go into the details of the second and third cases is that such mutants cannot survive under the conditions in question: any allele bearing a chance mutation of either type is immediately sieved from the population. And this should make sense: such mutants adjust their behavior away from the baseline in precisely the wrong contexts. The hawklike mutants, in case 2, switch to playing doves when encountering hawks. The dovelike mutants, in case 3, switch to being hawks when encountering hawks. Such mutant alleles end up being less fit than the hawk and dove alleles that gave rise to them.

We will now reexamine the first and fourth cases when it is somewhat costly to be a mutant. In particular, we imagine that the mutant's sensitivity to the presence or absence of the trait, and/or their propensity to change their behavior as a function of whether they sense that trait, however unconscious or instinctual, is slightly costly. While the semiotic capacity is usually beneficial, insofar as it makes their behavior flexible, it also requires extra resources, such that their fitness is slightly lowered. In effect, we look at scenarios that are identical to those analyzed above, except that the fitness of the mutant is adjusted as follows:

$$\overline{F}_g(m) = \overline{F}_g(A_3) = F_g(A_3) - k\,, \tag{5.23}$$

where $F_g(A_3)$ is given by equation (5.15) and $0 \leq k \leq 1$. We expect that mutants will be less robust as the value of k increases; we expect to recover the original scenarios (for case 1 and case 4) as $k \to 0$; and we expect to recover the classic scenario as $k \to 1$.

The fixed points of all three alleles turn out to be much more complicated when a cost is added. Rather than walk readers through the messy algebra, we will simply show the results.

5.6 The Fixed Points of Costly Agents

Just to be absolutely clear, the curves in all the graphs that follow show the fixed points of hawks, doves, and mutants as a function of v for various parameter values (within the range of possible values, given the scenarios in question). These are the points, in phase space, that the population evolves to (and then remain stably fixed), assuming that the system begins with an arbitrarily small initial frequency of mutants and a near-classic ratio of hawks to doves.

Figure 5.5 shows the fixed points of hawks, doves, and costly mutants when the mutants are hawklike (as per case 1). In particular, it shows such fixed points, as a function of v, when $e_1 = .5$ (and $e_2 = 1$), and costs range from $k = 0$ to $k = .06$. In all cases, $a_1 = 1 - a_2 = .9$ and $a_3 = a_1$.

Figure 5.6 shows the fixed points of hawks, doves, and costly mutants when the mutants are dovelike (as per case 4). In particular, it shows such fixed points, as a function of v, when $e_2 = .1$ (and $e_1 = 1$), and costs range from $k = 0$ to $k = .01$. In all cases, $a_1 = 1 - a_2 = .9$ and $a_3 = a_2$.

As may be seen through the changing graphs in figure 5.5 and figure 5.6, as the cost increases (from zero), the fixed points change from those investigated in sections 5.3 and 5.4 (in which being a hawklike mutant or a dovelike mutant came with no cost), to the classic case (in which mutants have been sieved from the population entirely, such that only hawks and doves remain).

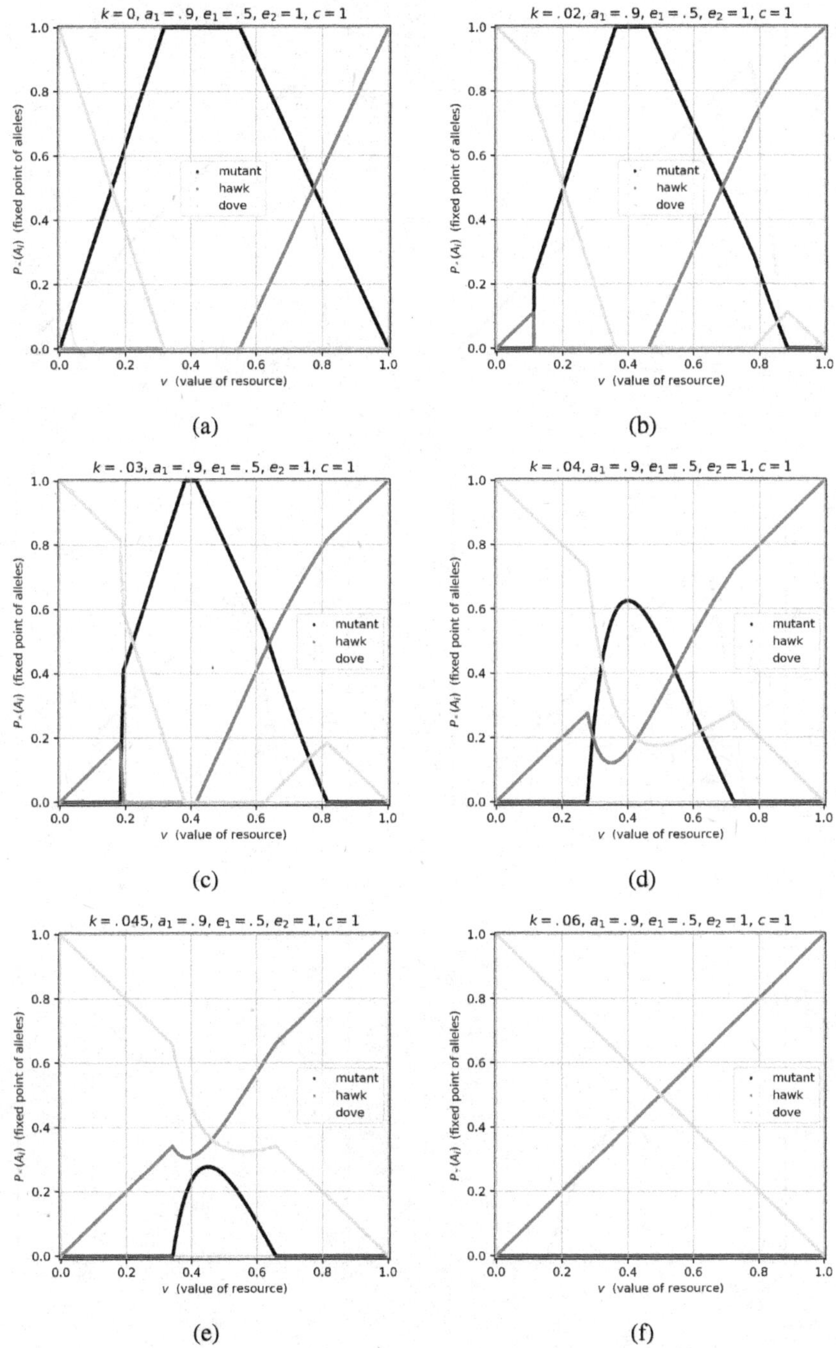

Figure 5.5: Fixed Points of Costly Hawklike Mutants

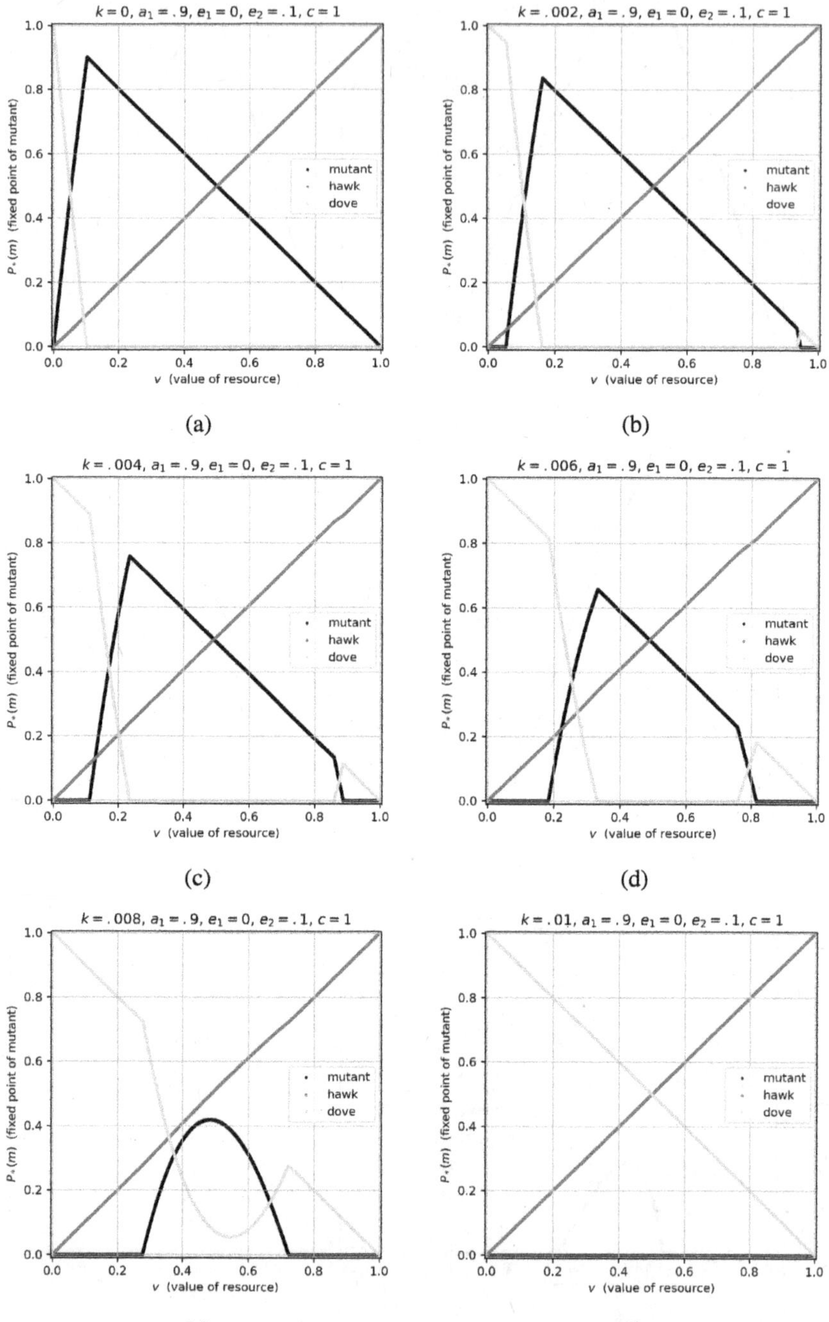

Figure 5.6: Fixed Points of Costly Dovelike Mutants

To interpret the graphs most easily, it is best to look at the x-axis of each one. Take figure 5.5(c), for example, which shows the fixed points of a hawklike mutant when $k = .03$, $e_1 = .5$ and $a_1 = .9$. In the leftmost and rightmost regions of the graph, corresponding to the lowest and highest values of v, the mutant cannot exist (the dark black line runs along the x-axis), so the hawk and dove take over the population completely. As may be seen, in both regions, the hawks and doves have fixed points that are identical to the classic scenario (for those values of v). Moving towards the center from both ends, there are now regions in which all three alleles coexist (i.e., there is no line along the x-axis). Moving towards the center some more, there is a region on the left where the hawk goes extinct (i.e., the gray line runs along the x-axis); and there is a region on the right where the dove goes extinct (i.e., the lightest line runs along the x-axis). Finally, there is a region near the center where both the hawk and dove alleles are sieved from the population and the mutant takes over completely (i.e., the gray line and the lightest line run along the x-axis, and the black line runs along the top of the graph).

Contrast the previous graph with the graph to the immediate right of it, such as figure 5.5(d), in which it is slightly more costly to have a semiotic capacity ($k = .04$), but all the other variables have identical values. As may be seen, there are now three regions: a leftmost region and a rightmost region, where the mutant cannot survive and hawks and doves take over the population completely (with their classic values); and a middle region where all three alleles can coexist, where the mutant's fixed-point frequency reaches its maximum.

Finally, as may be seen by looking across all the graphs, the more costly it is to have a semiotic capacity, the less well mutants do overall and the more mutants are constrained to the middle range of v values, where classic hawks and doves do about equally well.

This concludes Part II. We now turn from agents that evolve to agents that learn.

III AGENTS THAT LEARN

6 Reinforcement Learning

Overview

This chapter focuses on agents that learn from the results of their past behavior, and then change their future behavior accordingly. Section 6.1 discusses Edward Thorndike's "law of effect" and shows how it can be coupled to meaningful practices through the main equation (from chapter 1). Section 6.2 shifts the analysis from an individual agent to an ensemble of agents, and offers a difference equation for tracking the average behavior of agents within such an ensemble as they learn over time. Section 6.3 analyzes a scenario in which a signer and interpreter work together to learn a shared code, understood as an unmotivated set of conventions. Section 6.4 analyzes a scenario in which the code being learned is relatively motivated, insofar as one kind of sign is more expensive than the other, and insofar as one kind of object is more frequent than the other. Section 6.5 analyzes a scenario in which a parasite, or interloper, exploits the signer and interpreter's shared conventions, such that they must periodically establish a new set of conventions to thwart it. Finally, the conclusion considers the connection between movable fixed points and extractive social relations.

6.1 The Basic System

Suppose that an agent's behavior is guided by its embodied habits and its habits are determined by its past behavior—in particular, by the feedback that it received from the outcome of that behavior, in the guise of rewards or punishments, differential intensities of pleasure or pain, or simply positive or negative values. In short, the more a given behavior was rewarded in the past, the more likely it is to be performed in the present. This is the essence of Thorndike's law of effect (1898), which is sometimes described as "reinforcement learning."

This idea may be easily translated into our mathematical model of meaning. We will use the term *practice* to mean the instigation of a particular interpretant in the context of a particular sign, where such components are themselves causally coupled to particular objects and consequents in a given environment. We will assume that agents capable of reinforcement learning have a kind of embodied *register*, or set of habits, which we will denote as $W(I/S)$, understood as a matrix of weights indexed to particular practices (qua interpretant-sign pairs). The more an agent has been rewarded for a practice in the past, or positively sanctioned more generally, the greater the weight of that practice in its register; and the greater the present weight of a practice in an agent's register, the more likely the agent is to instigate that practice in the future. In this way, practices—understood as relations between signs and interpretants—are both the roots and fruits of registers; and registers themselves become differentially weighted over time as a function of the objects and consequents such practices relate to.

Note, then, that while traditional reinforcement learning is often framed as a kind of behavioralism, understood as dyadic relations between stimuli and responses, this retooling of the idea turns on thirdness (if not fourthness): the sign-interpretant relation only makes sense in relation to both the object-sign relation and the interpretant-consequent relation, and these relations only make sense in relation to the consequent-object relation (in relation to the interests, origins, and/or identity of the agent). Recall figure 1.2 from chapter 1. Something similar was also at work in chapter 3, as seen in our analysis of meaning in biosemiotic systems.

In short, just as an agent's register (understood as a kind of embodied interpretive ground, or habitus) influences its semiotic practices, an agent's semiotic practices influence its register. We will take up each direction of influence in turn.

To move from register to practice, we first need to move from an array of weights (as the trace of past experience) to an array of conditional probabilities (as propensities for future actions). While there are many ways to do this, we will assume that the agent's practices are determined by the current weights in its register as follows:

$$P(I_i/S_j) = \text{softmax}\big(W(I_i/S_j)\big) = \frac{e^{\beta \cdot W(I_i/S_j)}}{\sum_k e^{\beta \cdot W(I_k/S_j)}}. \tag{6.1}$$

In other words, after exponentiating all the entries in the matrix of weights, each of the columns in that matrix are normalized (by dividing each of the weights in a column by the sum of the weights in that column). As discussed in earlier chapters, this procedure is sometimes called the *softmax* function. The output of this function is thus a conditional probability distribution, $P(I/S)$, which constitutes the time-dependent *semiotic strategy* of the agent in question: namely, how likely they are to instigate a particular interpretant in the context of a particular sign (given their past experience in some environment). Here, variable β is a positively valued parameter that determines how concentrated the probability distribution is around larger weights in the register. As it increases, the probability that the agent instigates the most weighted interpretant (for a given sign) approaches 1. Phrased another way, the greater the value of β, the less likely the agent is to explore alternative practices.

We can now use the ground of the environment (given by the scenario in question), along with the current semiotic strategy of the agent (as just determined), to stochastically determine the next interaction the agent has with its environment. Assume, in particular, that a random number generator (or a set of biased coins) is used to generate objects, signs, interpretants, and consequents in accordance with such probability distributions (where the subscript g stands for "generated"):

$$P(O) \rightarrow O_g\,, \tag{6.2a}$$

$$P(S/O_g) \rightarrow S_g\,, \tag{6.2b}$$

$$P(I/S_g) \rightarrow I_g\,, \tag{6.2c}$$

$$P(C/I_g \wedge O_g) \rightarrow C_g\,, \tag{6.2d}$$

$$V(C_g \wedge I_g \wedge O_g) \rightarrow V_g\,. \tag{6.2e}$$

In other words, the priors are used to generate an object; the likelihoods, along with the generated object, are used to generate a sign; the agent's semiotic strategy, $P(I/S)$, along with the generated sign, are used to generate an interpretant; the eventualities, along with the generated interpretant and object, are used to generate a consequent; and the values, along with the generated consequent, interpretant and object, are used to determine the value thereby received by the agent.

To return to the ideas introduced at the end of chapter 2, it should be stressed that these values and probabilities, aside from $P(I/S)$, constitute the relatively objective ground of the environment, as opposed to the relatively subjective ground of the agent. That is, $P(O)$, $P(S/O)$, $P(C/I \wedge O)$, and $V(C \wedge I \wedge O)$ are determined by the actual causal patterns and valuable resources found in the environment (insofar as these resonate with the agent's sensations, instigations, capacities, and interests). The agent is, in effect, internalizing such patterns and resources as it interacts with the environment over time. The role of the main equation, in such a context, is not to tell the agent what it should do (given what the agent assumes will occur); rather, it specifies what will probabilistically occur in the environment in question, given what the agent does—which will, of course, end up transforming what the agent does, so far as that agent internalizes such patterns and resources in its embodied habits.

To find the weights in the agent's register in the next time slice, $W'(I/S)$, *all* the weights in the agent's current register, $W(I/S)$ are multiplied by a decay factor, $e^{-\alpha}$, and the weight of the agent's most recent practice (I_g in the context of S_g) is then increased by the value that was generated (V_g). More formally, the weights may be updated as follows:

$$W'(I/S) = W(I/S) \cdot e^{-\alpha}, \tag{6.3a}$$

$$W'(I_g/S_g) = W'(I_g/S_g) + V_g, \tag{6.3b}$$

where α is an adjustable parameter that determines how rapidly the effects of prior practices fade away from the agent's register, habitus, or memory. When $\alpha = 0$, for example, the effects of older practices never fade away, so older experiences are weighted just as strongly as newer experiences. For larger values of α, in contrast, the older an experience is, the less weight it is given in the agent's register, such that the present practice of the agent is determined more strongly by recent experiences. This ensures that the agent does not get stuck—at least too deeply—in the rut of past experience. So long as an environment is relatively stable, this parameter will usually be set to zero—insofar as past experiences, in such a static environment, are no less valuable than present experiences. As may be seen, and as should make sense, the larger the value obtained, the greater the change to the register (for the practice in question).

We now have the resources to track the non-deterministic behavior of an actual agent in a given environment. It is assumed, in particular, that an agent that has yet to experience an environment has all the weights in its register set to zero (which is equivalent to a blank slate), such that the agent is equally likely to instigate all possible interpretants in the context of any particular sign. As the agent experiences an environment by engaging in practices and receiving feedback in the form of positive and negative values, the weights in its register are updated accordingly. Note, then, that such a register is not only indexed to (or embodied in) a particular agent (whose experiences are themselves grounded in an environment), it is also indexed to a particular time, so it changes over time as the agent accrues more experience by interacting with its environment.

In what follows, as a means of assessing the rate at which, and the degree to which, an agent learns about its environment through the feedback it receives from the practices it undertakes, we will typically use *the expected value of its current semiotic strategy*, understood as

$$V\big(P(I/S)\big) = \sum_{jk} P(S_j) \cdot P(I_k/S_j) \cdot V(I_k/S_j)$$
$$= \sum_{ijk} P(O_i) \cdot P(S_j/O_i) \cdot P(I_k/S_j) \cdot V(I_k/S_j) \,. \qquad (6.4)$$

This is the value that the agent would receive, on average during an interaction with its environment, given its current semiotic strategy, $P(I/S)$, which is itself determined by the current register as per equation (6.1). As may be seen, the last term, $V(I_k/S_j)$, is determined by the main equation from chapter 1 (given the scenario in question). If the agent is indeed learning, the expected value of its semiotic strategy should—on average, and with certain caveats—increase over time until it reaches the optimal strategy for the scenario in question. In effect, the agent's subjective ground (as embodied in the weights of its register) comes into alignment with the relatively objective ground of the environment (understood as the actual causal patterns and valuable resources found therein).

Instead of creating a new scenario for this first example, we will borrow the scenario analyzed in chapter 2. But rather than track the behavior of a rational agent (that is deciding whether to take a medication given their symptoms), we are interested in tracking the behavior of a reinforced agent, and hence one that develops better health habits from past experience. While this is, to be sure, a somewhat strained scenario for this kind of learning, it allows us to compare the behavior of agents whose acts are more or less determined by reason or habit, and it should give readers an immediate sense of the dynamics without having to introduce all the specifics of a new scenario.

As may be seen in figure 6.1, which uses particular parameter values from the sickness and symptoms scenario, an agent that learns from experience on average engages in more and more valuable practices over time, approaching the semiotic strategy that a completely rational agent would adopt if it simply instigated the most valuable interpretant it could, for a given sign, as determined by the main equation. (Of course, this assumes that the rational agent's subjective ground was in accordance with the actual causal patterns and valuable resources of the environment in question.)

As may also be seen, the smaller learning parameter β is, the more slowly the learning process unfolds (on average). In particular, as we move from individuals with higher values of β to individuals with lower values of β, the more experience is required to learn the most valuable semiotic strategy, and thereby reach the largest expected value. That said, agents with large β are also more likely to undertake those practices with the highest weights in their register

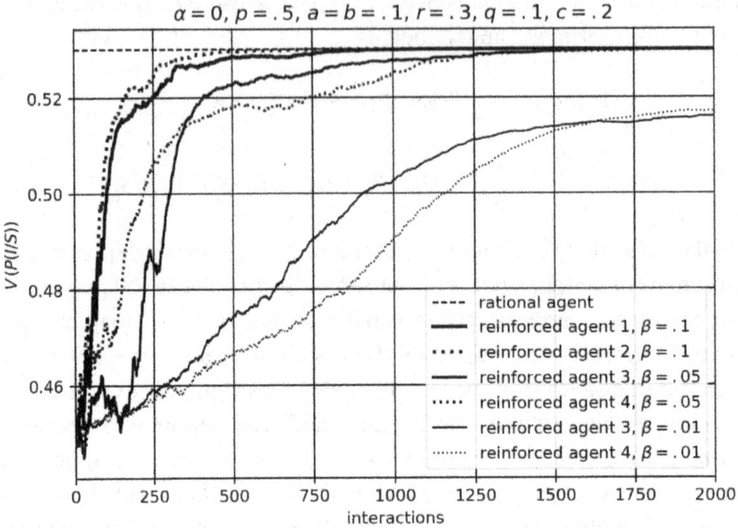

Figure 6.1: Expected Value of Individuals over Time

(as determined by their past experience, however serendipitous, as opposed to the facts on the ground), and so they can fail to fully explore, and ultimately benefit from, the space of possible practices.

The next section will examine the learning dynamics of an infinite population of such agents.

6.2 From Individual to Ensemble

Rather than track the actual practices of an individual agent, the main equation may be used to predict the *expected* practices of an *ensemble* of such agents (meaning the average behavior of an infinitely large population of such agents).

As before, it is assumed that the ground of the environment, in relation to the agent's capacities and interests, is specified such that the priors, likelihoods, eventualities, and values are known. Let $\overline{W}(I_i/S_j)$ denote the average weight of all the individual weights of the agents in this ensemble (for the practices in question) at a particular slice of time. To find the average or expected weight of the practices in the population during the next slice of time, $\overline{W}'(I_i/S_j)$, after all the agents in the ensemble have had one more interaction with their environment, simply multiply the current register by the decay factor, $e^{-\alpha}$, and then add the expected value of each practice in the ensemble to the current value. More formally, the following procedure is undertaken for each slice of time:

$$\begin{aligned}\overline{W}'(I_i/S_j) &= \overline{W}(I_i/S_j)\cdot e^{-\alpha} + \sum_k V(I_i/S_j)\cdot \overline{P}(I_i/S_j)\cdot P(S_j/O_k)\cdot P(O_k) \\ &= \overline{W}(I_i/S_j)\cdot e^{-\alpha} + V(I_i/S_j)\cdot \text{softmax}\left(\overline{W}(I_i/S_j)\right)\cdot P(S_j).\end{aligned} \tag{6.5}$$

As may be seen, $\text{softmax}\left(\overline{W}(I_i/S_j)\right)$ is equivalent to $\overline{P}(I_i/S_j)$, understood as the average or expected semiotic strategy of all the agents in the ensemble. $P(S_j)$, understood as the probability of encountering a particular sign during a time slice, may be found by summing over the product of the priors and likelihoods. And $V(I_i/S_j)$ is again the main equation from chapter 1.

It should be emphasized that equation (6.5) constitutes a *difference equation*: if the average weights of the practices of the population are known at some initial time, the average weights at any later time may be calculated. Indeed, using equation (6.4) with $\overline{P}(I_i/S_j)$ substituted for $P(I_i/S_j)$, the average expected value of the semiotic strategies of the agents in the ensemble may also be calculated. We will usually set the initial weights in the register of the ensemble to zero, or in certain cases, to small random numbers.

Let us return to the symptoms and sickness scenario introduced in chapter 2 and reused in section 6.1. The solid line in figure 6.2 shows the expected value, per time slice, for an ensemble of agents. It should be compared with the dotted lines, which show the expected value, per time slice, of the semiotic strategies of four individual agents, as calculated using the procedure discussed in section 6.1. (For the individual agents, as for the ensemble, $\beta = .1$, $\alpha = 0$, and the initial weights were set to zero.) The dashed line shows the expected value of a rational agent (whose subjective ground adequately represents the causal patterns and valuable resources in the environment), as modeled in chapter 2. Finally, the dash-dotted line shows the average expected value of 1,000 individual agents. As may be seen, the main equation, when used by a rational agent with perfect knowledge of its environment, calculates the limit point that equation (6.5) leads to in the long run. Phrased another way, relatively mindless agents, capable of semiotic reinforcement, act rationally on average in the long run. While the average of 1,000 agents is close to the ensemble of agents, it should approach it only in the limit of infinite agents. Therefore different runs will give slightly different lines for the average.

To be sure, we have been assuming that the environment is relatively fixed, such that an agent's weights can slowly accommodate to it. More interesting, arguably, is an environment that consists of other agents, that are themselves capable of reinforcement learning, such that an agent must accommodate its habits to an environment that is constituted by the practices—and hence habits—of other accommodating agents. We will tackle this issue in the

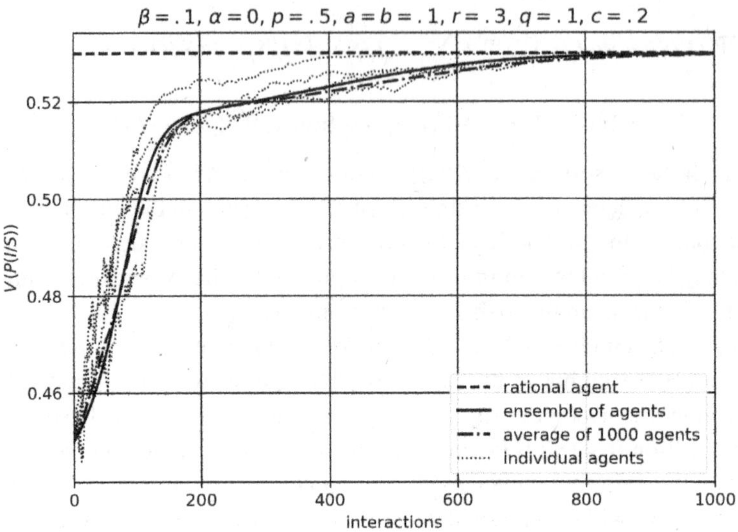

Figure 6.2: Expected Value of Ensemble over Time

next several sections, which will explore variations of a classic scenario: the creation of relatively motivated conventions.

6.3 Establishing a Convention

Suppose that there are two agents who must cooperate to secure a resource—say, some kind of tasty prey that they otherwise cannot catch on their own. One agent can see where the prey is located but not catch it; and the other agent can catch the prey but not see where it is located. The two agents therefore need to communicate to capture the prey (and share the value that it provides); but they must establish a code to do so. Recall figure 1.3, and the discussion around it. Luckily, both agents are capable of reinforcement learning, so they should be able to develop, or co-create, such a code after registering enough experience.

Assume, in particular, that the prey can be located in one of two places, which we may label O_1 and O_2. The first agent, upon perceiving the location of the prey, can express one of two signs, S_1 and S_2, such as a high-pitched shriek and a low-pitched grunt (or something else entirely). And the second agent, upon sensing the sign expressed by the first agent, can instigate one of two interpretants: I_1 is the action of going to O_1, and I_2 is the action of going to O_2. (In the interest of simplicity, consequents will be ignored; in effect, the interpreter always achieves whatever it attempts.) Assume as well that so long as the interpreter arrives at the place where the prey is located, it will definitely capture the prey and share its value with the signer, such that both agents will benefit equally. If, however, the interpreter goes to the wrong location, then neither agent benefits at all.

Crucially, whether the signer uses S_1 to indicate O_1 and S_2 to indicate O_2 (which might be called code 1), or vice versa (code 2), is immaterial; all that matters is that the interpreter engages in an analogous decoding, such that S_1 leads to I_1 and S_2 leads to I_2 (or vice versa). In other words, the success of the agents' behavior does not depend on which of the two codes they use; it only depends on their both using the same code. After translating the foregoing scenario into mathematical terms, we will track the behavior of the agents over time as they use reinforcement learning to establish such an otherwise arbitrary set of conventions.

Unlike the last example, in which there was only an interpreter, there is now both a signifying agent and an interpreting agent, where both such agents must have an embodied register to be capable of reinforcement learning. Let $W(S/O)$ denote the signer's register. Let $W(I/S)$ denote the interpreter's register. And assume that the value of the resource is v, such that the value that goes to either agent as a function of where the prey is located, O_1 versus O_2, and where the interpreting agent goes, I_1 versus I_2, is as follows:

$$V(I \wedge O) = \begin{bmatrix} v/2 & 0 \\ 0 & v/2 \end{bmatrix}. \tag{6.6}$$

In other words, both agents obtain half the value of the prey if the interpreter goes to the right location; and neither agent receives any value if the interpreter goes to the wrong location.

Assume that the registers of both agents are updated in the usual way. For the interpreter, equation (6.3) can be used, as described previously. For the signer, an analogous expression can be used:

$$W'(S/O) = W(S/O) \cdot e^{-\alpha}, \tag{6.7a}$$

$$W'(S_g/O_g) = W'(S_g/O_g) + V_g. \tag{6.7b}$$

Assume, finally, that the conditional probabilities of the agents' practices are determined from the weights in their registers in the usual way:

$$P(S_i/O_j) = \text{softmax}\big(W(S_i/O_j)\big) = \frac{e^{\beta \cdot W(S_i/O_j)}}{\sum_k e^{\beta \cdot W(S_k/O_j)}}, \tag{6.8a}$$

$$P(I_i/S_j) = \text{softmax}\big(W(I_i/S_j)\big) = \frac{e^{\beta \cdot W(I_i/S_j)}}{\sum_k e^{\beta \cdot W(I_k/S_j)}}. \tag{6.8b}$$

We now have the resources to track the non-deterministic behavior of the two coupled agents, where each agent constitutes a key part of the other agent's environment, such that each agent's agency is relatively distributed. As usual, we assume that the weights in the agents' registers are initially set to zero; and that we can use a random number generator (or a set of biased coins) to produce objects, signs, and interpretants in accordance with the foregoing probability distributions:

$$P(O) \to O_g, \tag{6.9a}$$

$$P(S/O_g) \to S_g, \tag{6.9b}$$

$$P(I/S_g) \to I_g, \tag{6.9c}$$

$$V(I_g \wedge O_g) \to V_g. \tag{6.9d}$$

In other words, the priors are used to generate an object; the signer's semiotic strategy, *P(S/O)*, along with the generated object, is used to generate a sign; the interpreter's semiotic strategy, *P(I/S)*, along with the generated sign, is used to generate an interpretant; and the values, along with the generated interpretant and object, are used to determine the value thereby generated.

We then add this value to each of the agents' registers, using equations (6.3) and (6.7), indexed to the actual practice that it performed (be it a sign in the context of an object or an interpretant in the context of a sign), update their semiotic strategies accordingly, using equation (6.8), and repeat the same sequence of steps for the next interaction (or slice of time). And so it continues indefinitely.

As usual, we want to plot the success rate of such agents, as a function of time slice, so we can see how fast and how well they establish a common code. To do this, it is easiest to plot the expected value of their semiotic strategies:

$$V\big(P(I/S)\big) = V\big(P(S/O)\big) = \sum_{ijk} V(I_k \wedge O_i) \cdot P(I_k/S_j) \cdot P(S_j/O_i) \cdot P(O_i). \tag{6.10}$$

Note that the expected value of the signer's semiotic strategy, at any point in time, is equal to the expected value of the interpreter's semiotic strategy. This makes sense given the details of this scenario: the fates of these otherwise independent agents are symmetrically entangled.

Figure 6.3 plots the expected value of the semiotic strategies of four individual pairs of signers and interpreters as a function of time (understood as the number of interactions that they have had with each other). Here, for the sake of simplicity, we have set $p = .5$ (such that the prey is equally likely to be found in either location).

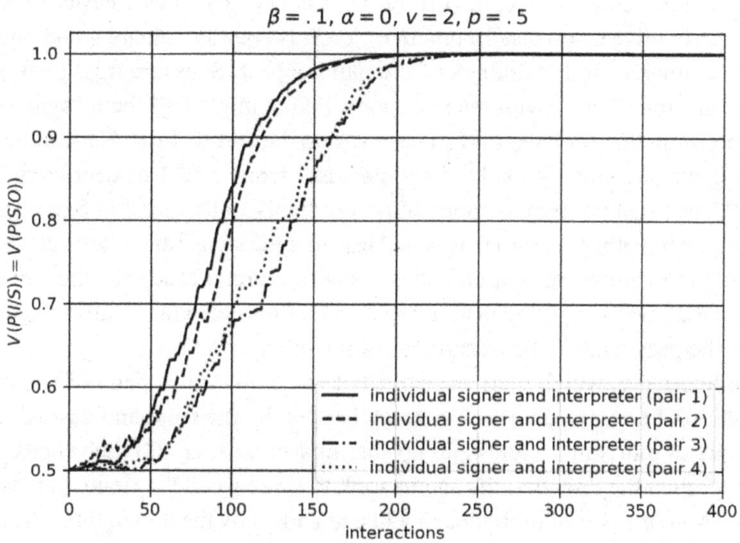

Figure 6.3: Establishing a Code

As may be seen, whereas all such pairs of agents begin with an expected value of .5 (corresponding to how well each agent does if both agents behave randomly), they all converge on an expected value of 1 (such that they end up correctly coordinating their behavior in each time slice, and so they always receive half the value of the resource, where $v = 2$).

Such perfect coordination was achieved because all four individual pairs of interacting agents ended up establishing a code through reinforcement learning (at the end of 200 interactions or so), such that the signers and interpreters were perfectly in sync with each other. In this particular run, pairs (1) and (3) ended up converging on code 1 (where S_1 stands for O_1, giving rise to I_1; and S_2 stands for O_2, giving rise to I_2). Meanwhile, pairs (2) and (4) ended up converging on code 2 (where S_1 stands for O_2, giving rise to I_2; and S_2 stands for O_1, giving rise to I_1).

To see this convergence more clearly, figure 6.4 plots $P(S_1/O_1)$ and $P(S_1/O_2)$ for individual signers and $P(I_1/S_1)$ and $P(I_1/S_2)$ for individual interpreters. Figure 6.4a does this for the signer and interpreter in pair (1) of the last graph; and figure 6.4b does this for the signer and interpreter in pair (2) of that graph. As may be seen, whereas the first pair of agents converged on code 1, the second pair of agents converged on code 2. Given the way that the scenario was set up, neither code is favored (insofar as it is cheaper to use, easier to learn, or more effective or precise in practice): each is equally valuable, and equally likely, so long as both members of the pair adopt it. Saussure might call such codes "arbitrary"; and, with many caveats, Peirce might call them "symbolic."

Everything that we have said so far turns on the fact that $p = .5$, meaning that both objects are equally likely. As p increases from .5 to 1 or decreases from .5 to 0, such that the prey is more and more likely to be found in one position rather than the other, it matters less and less that the signer and interpreter share a code. For example, as p approaches 1, the interpreter (and thus the signer as well) can do fine simply by instigating I_1 each time, such that it always goes to where the prey tends to be, regardless of the other agent's sign.

See figure 6.5, which plots the expected value of various pairs of agents, as a function of time, when $p = .7$. As may be seen by the solid and dashed lines, two pairs of individual agents end up sharing a code after 400 time slices. This is longer than the time that the agents took to share a code in figure 6.2 (where $p = .5$). And one pair of individual agents, as shown by the dotted line, got stuck in a lower-value solution—one in which the interpreter instigated I_1 regardless of the sign, and in which the signer always expressed S_2 when the object was O_1, but otherwise behaved relatively randomly.

Figure 6.6 attempts to quantify and visualize such tendencies. In particular, we tracked 1,000 runs of pairs of interacting agents, each with 1,000 time slices, for various values of p between 0 and 1. At the end of each run, it was determined whether the agents ended up using code 1 or code 2 (the optimal solution) or whether the interpreter simply instigated the interpretant most likely to overlap with the object (a somewhat satisfactory solution when p is near zero or 1). Note that the frequencies of the four possible outcomes do not always add up to 1. This is because, even after 1,000 slices of time, the agents may not yet have ended up in a stable solution, especially for extreme values of p.

Loosely, the greater the asymmetry in the position of the object (and hence the more predictable the environment is, such that the less informative or valuable the sign is), the longer it takes for the agents to establish a code and the

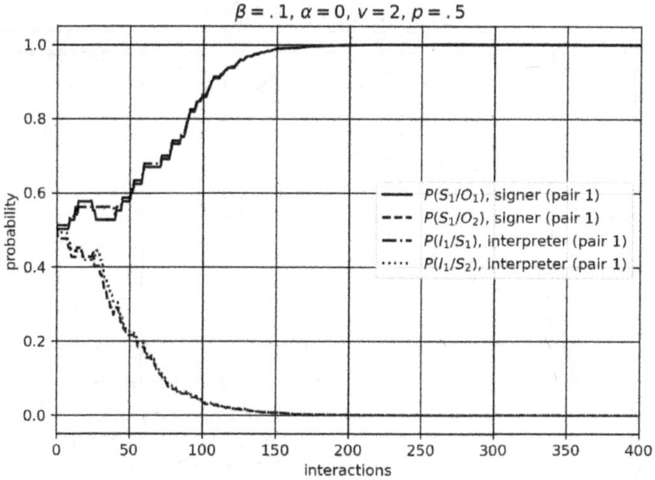

(a)

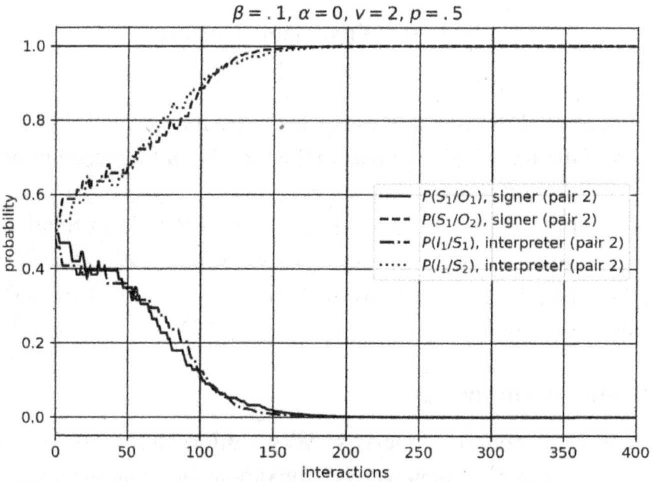

(b)

Figure 6.4: Converging on a Common Code

more likely that the interpreter develops a shortcut instead—a shortcut that might always be altered if more time to learn is added. Indeed, not withstanding the effects of an asymmetric environment (in which one kind of object is more probable than another), the tendency of the agents to establish a common code, or set of conventions, through reinforcement learning is incredibly robust

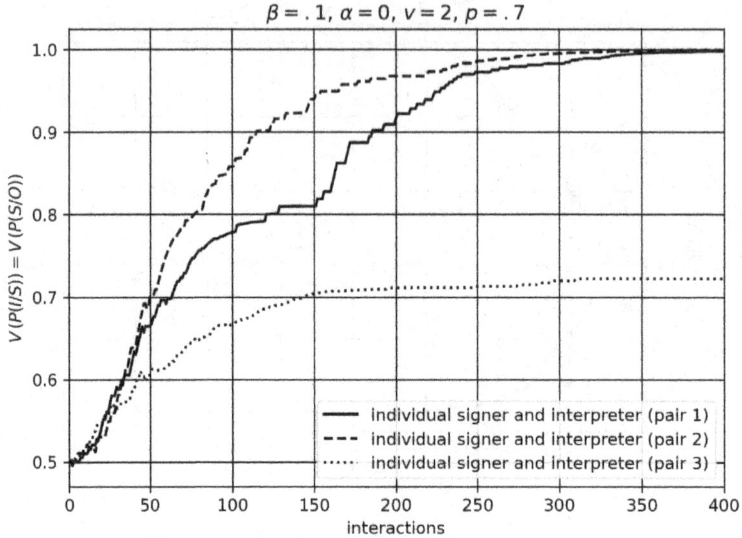

Figure 6.5: Predictable Objects

(as may be seen by the broad flatness of the solid and dashed lines). Note that the jerkiness of the plots is due in part to the fact that p increased in increments of .05.

We now examine the establishment of a common code in relatively asymmetric environments, when the semiotic processes in question have differential costs and benefits. In so doing, we will make the move from arbitrary to motivated conventions.

6.4 Motivating Conventions

We now analyze a scenario, otherwise very similar to the last one, in which one sign is slightly more expensive than the other, such that the signer receives a smaller payoff when it encodes the position of the prey using that sign. It may be expensive for a variety of reasons that need not be specified (say that it takes more energy to produce or takes more time to utter). As will be seen, the conventions established by the agents are now relatively motivated: the more frequent object tends to be encoded using the less expensive sign.

All the mathematical details from section 6.3 carry over to this scenario, except for the following changes: First, the payoff to the signer and interpreter now depends on the sign (in addition to the object and interpretant). For the

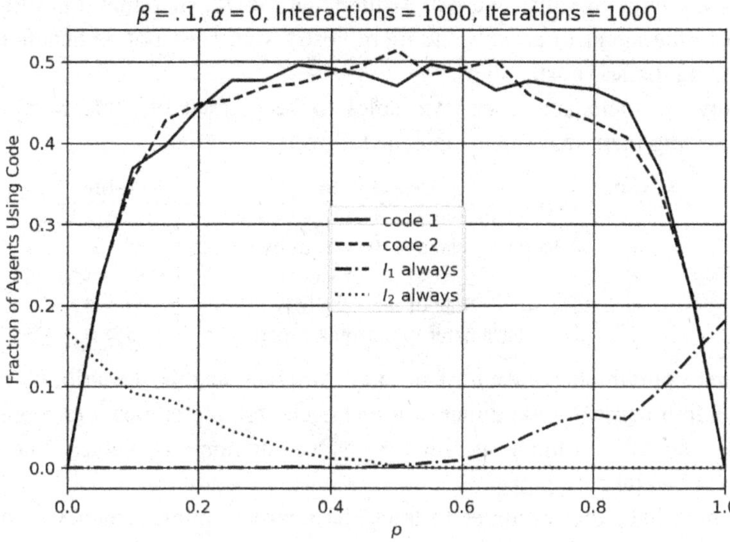

Figure 6.6: Relative Predictability

interpreter, the change in the matrix of values is trivial (insofar as it is still independent of the sign):

$$V_I(I \wedge S_1 \wedge O) = \begin{bmatrix} v/2 & 0 \\ 0 & v/2 \end{bmatrix}, \tag{6.11a}$$

$$V_I(I \wedge S_2 \wedge O) = \begin{bmatrix} v/2 & 0 \\ 0 & v/2 \end{bmatrix}. \tag{6.11b}$$

For the signer, however, the payoff now depends on whether S_1 or S_2 was used to encode the position of the object:

$$V_S(I \wedge S_1 \wedge O) = \begin{bmatrix} v/2 - c & -c \\ -c & v/2 - c \end{bmatrix}, \tag{6.12a}$$

$$V_S(I \wedge S_2 \wedge O) = \begin{bmatrix} v/2 & 0 \\ 0 & v/2 \end{bmatrix}. \tag{6.12b}$$

As may be seen, we have made S_1 more costly to the signer than S_2, by c. In particular, every time the signer uses S_1 to encode the position of the object, regardless of the response of the interpreter or the success of their interaction, they lose c units of value. We will assume that $c < v/2$, such that it is still beneficial to the agent to employ the more costly sign (just not as beneficial as employing the less costly sign).

Insofar as there are several variables to keep track of, it is worthwhile summarizing them in a single table, as follows:

Variable	**Description**	**Value**
p	Probability resource is located at O_1	$0 \leq p \leq 1$
v	Value of resource to be split by agents	$v = 2$
c	Cost of S_1 for signer	$0 \leq c < v/2$
α	Rate of memory loss	$\alpha \geq 0$
β	Parameter of softmax function	$\beta \geq 0$

The first column shows each of the key variables in this scenario. The second column describes the dimension that each variable is meant to represent. Finally, the third column specifies the value, or range of values, that such variables can take.

We now have the resources to track the non-deterministic behavior of the two agents. Assume, as usual, that the weights in the agents' registers are initially set to zero; and assume that a random number generator can be used to produce objects, signs, and interpretants in accordance with the foregoing probability distributions. We can then iterate through the following steps:

$$P(O) \rightarrow O_g \,, \tag{6.13a}$$

$$P(S/O_g) \rightarrow S_g \,, \tag{6.13b}$$

$$P(I/S_g) \rightarrow I_g \,, \tag{6.13c}$$

$$V_S(I_g \wedge S_g \wedge O_g) \rightarrow V_g(signer) \,, \tag{6.13d}$$

$$V_I(I_g \wedge S_g \wedge O_g) \rightarrow V_g(interpreter) \,. \tag{6.13e}$$

In other words, the priors are used to generate an object; the signer's semiotic strategy, $P(S/O)$, along with the generated object, are used to generate a sign; the interpreter's semiotic strategy, $P(I/S)$, along with the generated sign, are used to generate an interpretant; and the values of the signer and interpreter, along with the generated interpretant, sign, and object, are used to determine the values that are received by the signer and interpreter, respectively.

We then add these values to each of the agents' registers using equations (6.7) and (6.3), indexed to the actual behavior that they each performed (be it a sign in the context of an object or an interpretant in the context of a sign), update their semiotic strategies accordingly using equation (6.8), and repeat the same sequence of steps for the next interaction (or slice of time). And so it continues indefinitely.

Finally, we need to slightly adjust our equations for calculating the expected value that goes to the signer and interpreter, respectively, as a function of their current semiotic strategies:

$$V\big(P(S/O)\big) = \sum_{ijk} V_S(I_k \wedge S_i \wedge O_j) \cdot P(I_k/S_i) \cdot P(S_i/O_j) \cdot P(O_j), \qquad (6.14a)$$

$$V\big(P(I/S)\big) = \sum_{ijk} V_I(I_i \wedge S_j \wedge O_k) \cdot P(I_i/S_j) \cdot P(S_j/O_k) \cdot P(O_k). \qquad (6.14b)$$

As may be seen, in contrast to the previous scenario, the expected values of the signer and interpreter are no longer the same. This reflects the fact that signers pay a price when S_1 is used, but not interpreters.

Figure 6.7 plots the expected value, as a function of time slice, for four pairs of signers and interpreters. Each pair is denoted with a different pattern (solid, dashed, dash-dotted, and dotted); and the ultimately higher line in each pattern corresponds to the interpreter, whereas the ultimately lower line in each pattern corresponds to the signer. In this run, the cost of S_1 was relatively low ($c = .1$), and the prey was more likely to be be found at O_2 than O_1 ($p = .3$). As may be seen, all four pairs of agents learned a shared code. Three pairs of agents ended up converging on code 1 (pairs 1, 3, and 4); and one pair of agents ended up converging on code 2 (pair 2). In other words, three pairs of agents ended up converging on the better of the two codes for the parameters in question: for the more expensive sign is best used to encode the least frequent object. As should be expected, the expected value of the signer in each pair is lower than the expected value of the interpreter—and more so for the second pair, precisely because that pair of agents settled on a code in which the more expensive sign was used to encode the more frequent object.

The graphs in figure 6.8 show the relative frequency with which codes 1 and 2 were adopted by 100 pairs of agents (after 1,000 interactions with each other), as a function of p, for various values of c. As may be seen, so long as S_1 is somewhat costly ($c > 0$), code 1 tends to be adopted when $p > 1/2$, and code 2 tends to be adopted when $p < 1/2$. Agents tended to converge on the most cost-effective code. In effect, the less costly sign (which is also the smaller, or less marked, sign) becomes the more frequently used sign, insofar as it comes to stand for the more frequently occurring object.

As may be seen, when $c = 0$, meaning that neither sign was more costly than the other, both codes were equally likely to be adopted in the population for most values of p. At the end points, however, when $p < .1$ or $p > .9$, neither code ended up being adopted, as it was more useful for the interpreter to always instigate I_1 or always instigate I_2. This issue was discussed in section 6.3. As the value of c increases, from 0 to .2 and from .2 to .4, code 1 tends

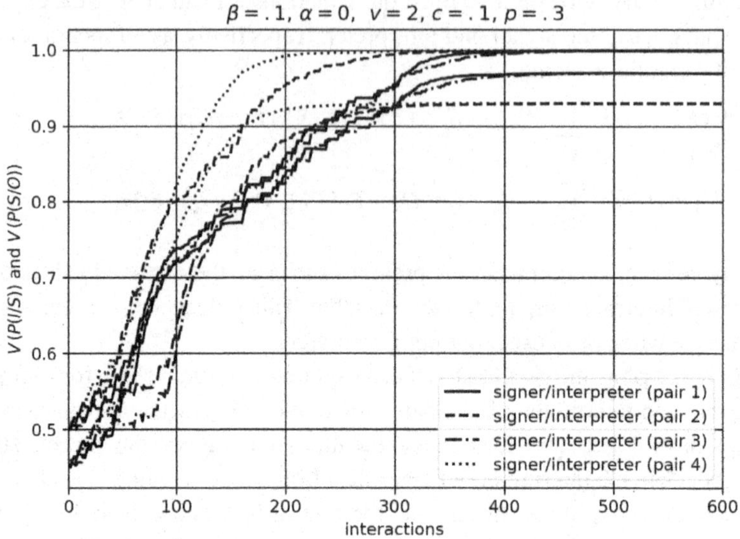

Figure 6.7: Motivated Conventions and Time-to-Code

to be used when $p < .5$ and code 2 tends to be used when $p > .5$. This double-peaked pattern continues for even more costly signs ($c = .8$); however, as may be seen by the relative shortness of the peaks, fewer pairs of agents managed to settle on code 1 or code 2, or some other encoding, as the value generated by the semiotic interaction was very small for the signer (and so its learning rate was correspondingly slow). Note that the graphs are relatively rough, in part because the value of p was increased in increments of .05, and in part because the values come from a population with only 100 members (so the average can vary).

Such results should not be too surprising, to be sure. More interesting is a scenario in which there is a third semiotic agent, understood as a parasite, that can exploit the conventionality of the code employed by the signer and interpreter. As will be shown in the next section, such a scenario can lead to cyclic behavior where the signer and interpreter have to keep changing their code to stay ahead of the parasite.

6.5 Parasites

We now introduce a third agent to our scenario—one that is able to learn but unwilling to share. Like the interpreter, it too can perceive the sign and, through experience, learn whether it is best to go to one location or the other to secure

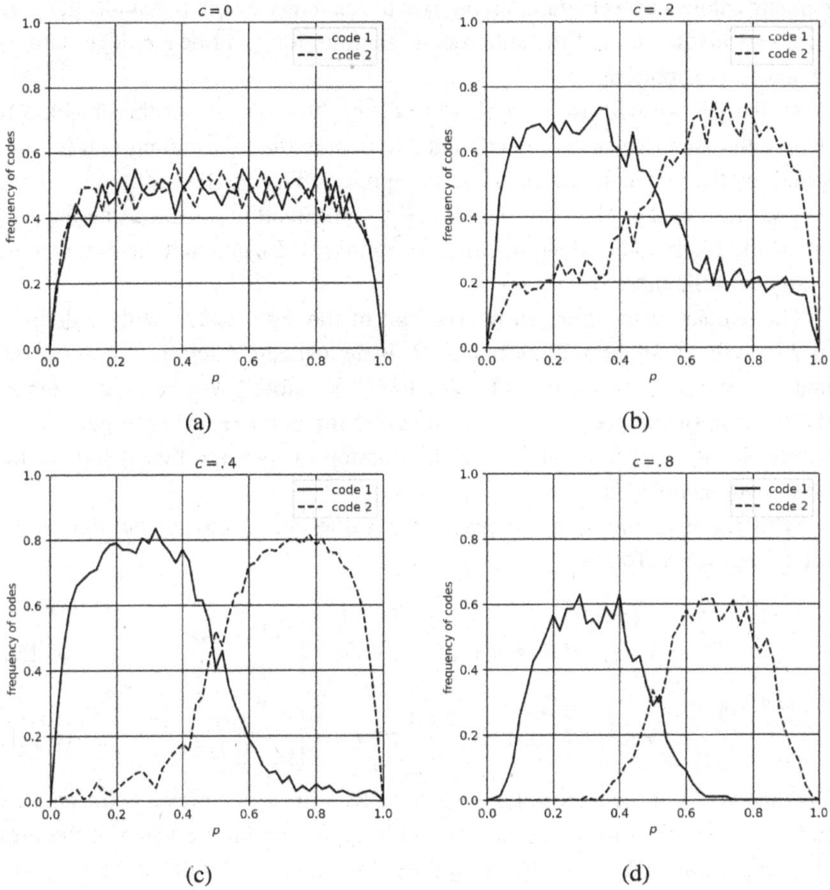

Figure 6.8: Motivated Conventions and Costly Signs

the prey. However, unlike the interpreter, it uses this information to locate the prey before the other two agents arrive, such that it can keep most of the prey's value for itself. We will label this third agent the *parasite*—that which takes without giving. In effect, it exploits the code that is established by the other two agents, using their shared means of communication for selfish ends.

Crucially, the more the parasite can benefit from that code, the less that code is beneficial to the other two agents. As will be seen, this often causes the signer and interpreter to find another code and to use this new code until the parasite catches up yet again, giving rise to relatively cyclic behavior: the periodic establishment, and the subsequent exploitation (and hence degradation), of shared conventions.

As before, we will use variable O to denote the location of the object, S to denote the sign that is expressed, and I to denote the interpretant that is instigated (by the original, non-parasitic interpreter). That is, the action of going to one location or the other. For the sake of simplicity, it is assumed that the prey is equally likely to be found in either location ($p = .5$), and neither sign is more costly than the other ($c = 0$).

The parasite also offers an interpretant of the sign, which will be denoted as J to avoid confusion. In particular, J_1 is the parasite's action of going to O_1 and J_2 is the parasite's action of going to O_2. Variable f will be used to denote the fraction of the prey that is consumed by the parasite (when it goes to the correct location), such that $1-f$ is the fraction of the prey that is left for the other two agents to share.

With these assumptions in place, the values received by the signer and interpreter are as follows:

$$V_S(J_1 \wedge I \wedge O) = V_I(J_1 \wedge I \wedge O) = \begin{bmatrix} \frac{(1-f)\cdot v}{2} & 0 \\ 0 & \frac{v}{2} \end{bmatrix}, \tag{6.15a}$$

$$V_S(J_2 \wedge I \wedge O) = V_I(J_2 \wedge I \wedge O) = \begin{bmatrix} \frac{v}{2} & 0 \\ 0 & \frac{(1-f)\cdot v}{2} \end{bmatrix}. \tag{6.15b}$$

In other words, if the interpreter goes to the location of the prey and the parasite did not already get there, the signer and interpreter split the value of the prey (which is v, as before). If the interpreter goes to the location of the prey and the parasite has already been there, the signer and interpreter split the value of the remains of the prey, meaning that fraction $(1-f)\cdot v$ that was left by the predator. Finally, if the interpreter goes to the wrong location, then neither the signer nor the interpreter receives any value (and this is true regardless of the parasite's behavior).

The values of the parasite are complementary:

$$V_P(J_1 \wedge I \wedge O) = \begin{bmatrix} f\cdot v & 0 \\ f\cdot v & 0 \end{bmatrix}, \tag{6.16a}$$

$$V_P(J_2 \wedge I \wedge O) = \begin{bmatrix} 0 & f\cdot v \\ 0 & f\cdot v \end{bmatrix}. \tag{6.16b}$$

In other words, if the parasite goes to the location of the prey, it receives some fraction of the prey's value ($f \cdot v$), where $0 \leq f \leq 1$. And if the prey goes to the wrong location, it does not receive any value.

We now have three sets of weights, or embodied registers, to keep track of: those belonging to the signer, $W(S/O)$; those belonging to the interpreter, $W(I/S)$; and those belonging to the parasite, $W(J/S)$. As before, all of these can be converted to conditional probabilities, which constitute the semiotic strategies of the three agents, using the softmax function:

$$P(S/O) = \text{softmax}\big(W(S/O)\big)\,, \tag{6.17a}$$

$$P(I/S) = \text{softmax}\big(W(I/S)\big)\,, \tag{6.17b}$$

$$P(J/S) = \text{softmax}\big(W(J/S)\big)\,. \tag{6.17c}$$

As usual, it is assumed that the weights in the agents' registers are initially set to zero; and a random number generator is used to produce objects, signs, and interpretants in accordance with the foregoing probability distributions. We can then step through the following procedure for every slice of time:

$$P(O) \rightarrow O_g\,, \tag{6.18a}$$

$$P(S/O_g) \rightarrow S_g\,, \tag{6.18b}$$

$$P(I/S_g) \rightarrow I_g\,, \tag{6.18c}$$

$$P(J/S_g) \rightarrow J_g\,, \tag{6.18d}$$

$$V_S(J_g \wedge I_g \wedge O_g) \rightarrow V_g(signer)\,, \tag{6.18e}$$

$$V_I(J_g \wedge I_g \wedge O_g) \rightarrow V_g(interpreter)\,, \tag{6.18f}$$

$$V_P(J_g \wedge I_g \wedge O_g) \rightarrow V_g(parasite)\,. \tag{6.18g}$$

In other words, the priors are used to generate an object; the signer's semiotic strategy, $P(S/O)$, along with the generated object, is used to generate a sign; the interpreter's semiotic strategy, $P(I/S)$, along with the generated sign, is used to generate its interpretant (I); the parasite's semiotic strategy, $P(J/S)$, along with the generated sign, is used to generate its interpretant (J); and the values of all three agents, along with the two generated interpretants and the object, are used to determine the value that is received by each of the agents. These values are added to the weights in each agent's register in the usual way. For example, in the case of the parasite, we have

$$W'(J/S) = W(J/S) \cdot e^{-\alpha}, \tag{6.19a}$$

$$W'(J_g/S_g) = W'(J_g/S_g) + V_g(parasite). \tag{6.19b}$$

And it is similar for the other two agents. Crucially, insofar as all three agents are trying to learn through experience in a relatively labile environment, it is useful for them to forget older experiences and focus more on newer ones. Accordingly, we will choose a value for the memory-decay parameter, α, that is greater than zero.

Finally, we can calculate the expected value of each agent's semiotic strategy, at any point in time, using the following equations:

$$V\big(P(S/O)\big) = \sum_{ijkl} P(O_i) \cdot P(S_j/O_i) \cdot P(I_k/S_j) \cdot P(J_l/S_j) \cdot V_S(J_l \wedge I_k \wedge O_i), \tag{6.20a}$$

$$V\big(P(I/S)\big) = \sum_{ijkl} P(O_i) \cdot P(S_j/O_i) \cdot P(I_k/S_j) \cdot P(J_l/S_j) \cdot V_I(J_l \wedge I_k \wedge O_i), \tag{6.20b}$$

$$V\big(P(J/S)\big) = \sum_{ijkl} P(O_i) \cdot P(S_j/O_i) \cdot P(I_k/S_j) \cdot P(J_l/S_j) \cdot V_P(J_l \wedge I_k \wedge O_i). \tag{6.20c}$$

As may be seen, except for the rightmost factor, which denotes the value received by the agent in question, all these equations are identical. This reflects the fact that the value received by each agent is determined not just by its own behavior, but also by the behavior of the other two agents.

When we analyze the semiotic strategies and expected values of the three agents in this scenario, for different values of f, we find the following patterns. First, when $f = 0$, the behaviors of the signer and interpreter are identical to those of their counterparts in section 6.3: they quickly adopt either code 1 or code 2, and then use that code to communicate ever after, each with an expected value of 1. The parasite, not garnering any value from the prey whether it goes to the right location or not, and therefore not benefiting from their communication, ends up behaving randomly, with an expected value of zero.

Second, as f increases towards 1/2, such that the parasite benefits more and more from the prey (but not too much), the signer and interpreter continue to adopt one code or the other, and the parasite comes to adopt whatever code they end up using. As such, the parasite benefits from their communication, but not enough to disrupt their usual practices. And just as the expected value of the parasite gets larger and larger (moving from 0 to 1 as f approaches 1/2), insofar as it is taking a larger fraction of the prey, the expected values of the signer and interpreter get smaller and smaller (moving from 1 to 1/2). As may be seen in figure 6.9, for example, as f moves from .1 to .4, the expected values of the signer and interpreter decrease, while the expected value of the parasite increases. As may be seen in figure 6.10, the signer and interpreter settle on one code or the other, and the parasite follows suit. For small f, it approaches their code (but is still relatively random); for large f, in contrast, it adopts their code almost completely. In all cases, the final behavior (after around 600 interactions or time slices), would continue indefinitely (with minor fluctuations) once it is reached. (As in earlier versions of this scenario, $v = 2$ and $\beta = .1$. Unlike those earlier versions, however, $\alpha = .01$, so that the agents tend to forget—if only very slowly—what they previously learned.)

Third, as f increases from 1/2 to 1, such that the parasite is taking more than half of the prey's value for itself, a qualitatively new pattern of behavior arises. The signer and interpreter begin to alternate between code 1 and code 2, as does the parasite—which is always a step behind them. In particular, as the signer and interpreter establish one code or the other, their expected value increases; however, the parasite soon catches on to their code and starts taking most of the prey for itself, such that their expected value decreases; so they switch to the other code, and it soon follows suit. As such, the behavior of all three agents is periodic, with the period decreasing as f increases. Moreover, the maximum amplitude of the parasite's expected value increases as f increases, while the minimum amplitude of the signer and interpreter's expected value decreases. Figure 6.11 shows the expected value of the three agents, as a function of time, for four values of f; and figure 6.12 shows the semiotic strategies of the three agents for two of these values. As may also be seen, especially when $f = .6$, while the signer and interpreter have sharp highs and flat lows, the parasite has relatively fast increases and slow decreases.

Finally, when $f \geq .9$, the periodic behavior becomes much less frequent and a more static behavior emerges. This behavior can be glimpsed in figure 6.11d, after around 3,000 slices of time, where the expected values of the three agents stay fixed. If we look at the semiotic strategies of these agents in that region of time, as shown in figures 6.12c and 6.12d, we see that the parasite ends up doing J_2 regardless of the sign, while the signer and interpreter adopt the same code when the prey is located at O_1, but behave randomly when the prey is located at O_2. In effect, the parasite takes enough of the prey for itself, and the signer and interpreter have so little left, that behaving in a rigid or random way is good enough.

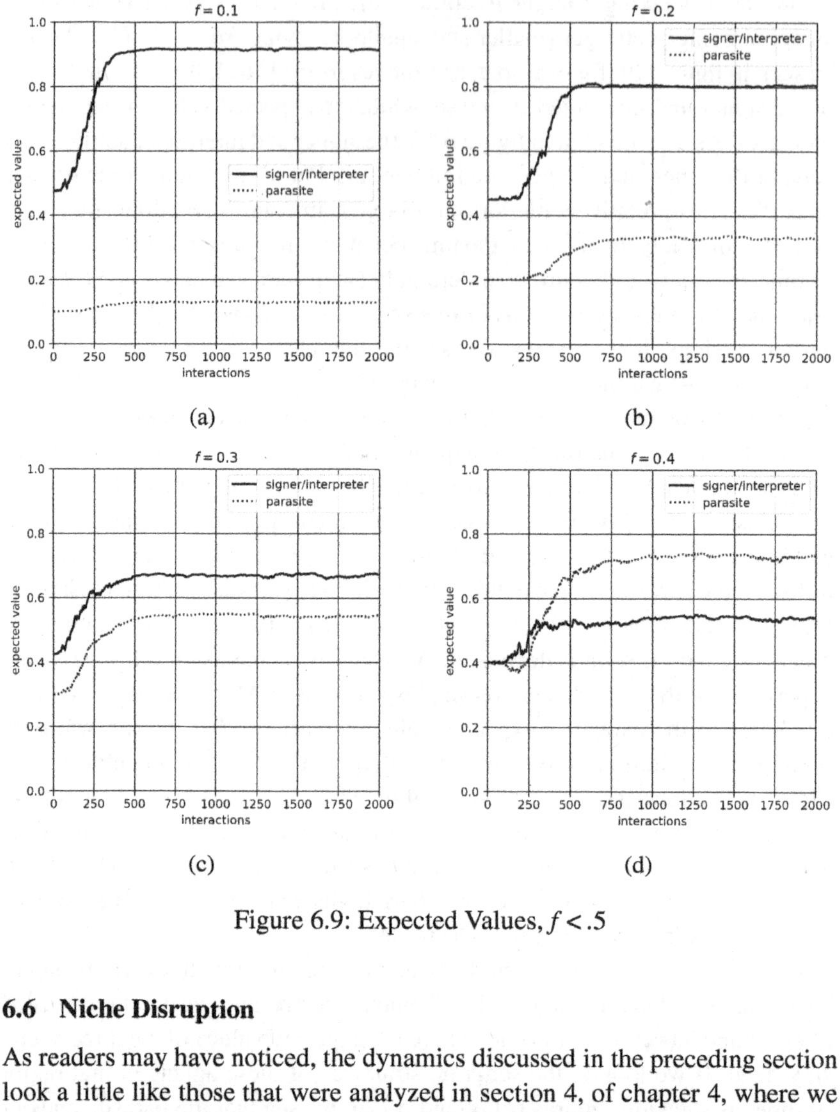

Figure 6.9: Expected Values, $f < .5$

6.6 Niche Disruption

As readers may have noticed, the dynamics discussed in the preceding section look a little like those that were analyzed in section 4, of chapter 4, where we took up the competitive relation between plants (as signers) and animals (as interpreters), as well as the oscillatory dynamics of their evolutionary trajectories. But rather than examine a confrontation between two agents, each of whom benefits at the expense of the other, we examined a relation among three agents, one of whom benefits at the expense of the other two. This meant that the code created by the signer and interpreter was designed not just to allow communication between them, but also to prohibit communication with the

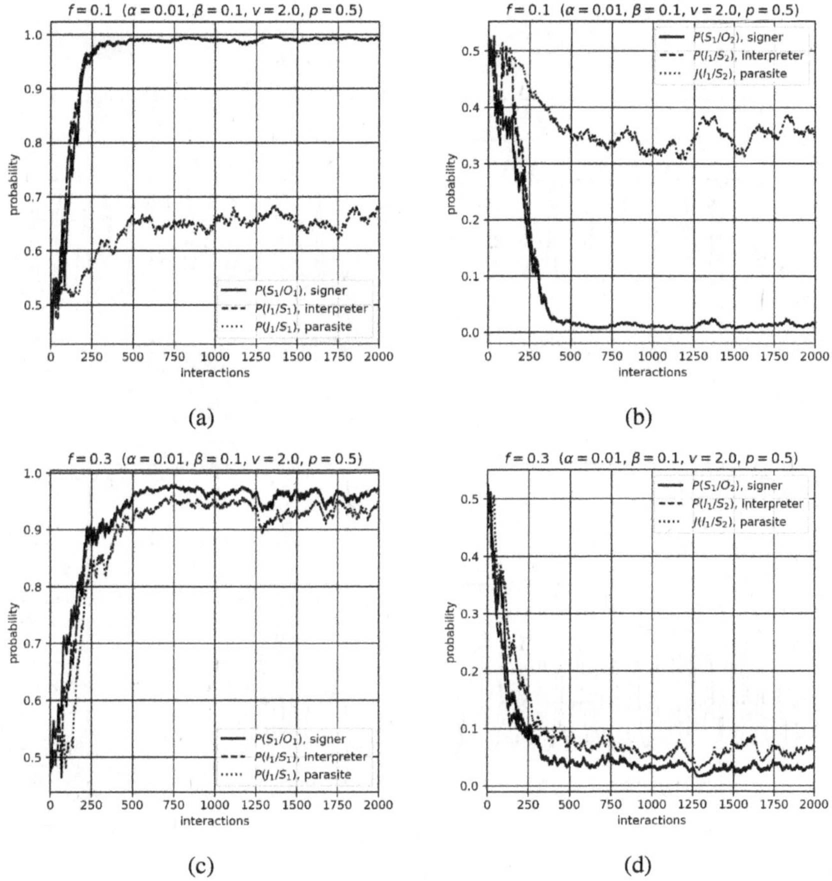

Figure 6.10: Semiotic Strategies, $f < .5$

parasite. Phrased in another way, the features of the code and, in particular, its temporal transformations, indexed not only a relatively amicable social relation between signer and interpreter, but also a relatively extractive, or despicable, social relation between the parasite and those two agents. Codes, and interpretive grounds more generally, function as thorns and camouflage no less than bridges and doors.

To be sure, by the word "designed" we do not mean intentionally constructed, or even evolutionarily selected. We simply mean historically reinforced—or regimented—by past interactions, and their outcomes, as embodied in the habits, or registers, of the interactants. And like the example of plants and animals, our discussion of parasites also highlighted the fact that

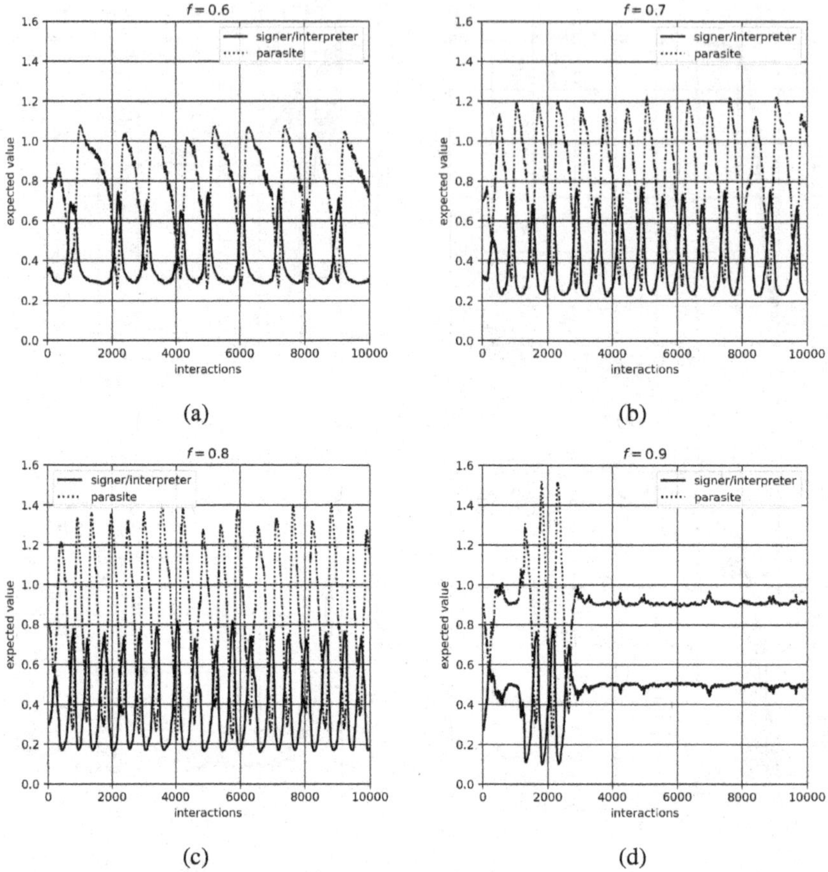

Figure 6.11: Expected Values, $f > .5$

a key part of any organism's environment is other organisms, themselves learning and evolving, such that there is no fixed fixed point, so to speak, to evolve to: each time the organisms change in light of their environment, the environment changes, thereby shifting the direction of change once again. Such niches are disrupted as soon as—or at least soon after—they are constructed.

This concludes our discussion of relatively rudimentary agents who learn by reinforcement learning. We now turn to agents whose interpretive grounds are embodied in neural networks, as opposed to relatively simplistic registers, and whose capacities to learn, and potential semiotic complexity more generally, are far greater.

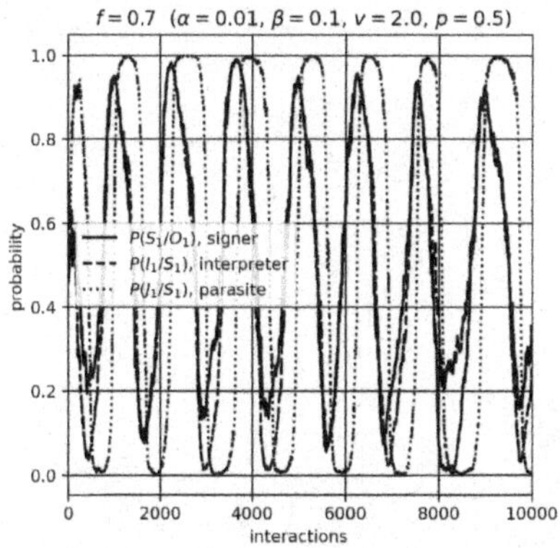

(a)

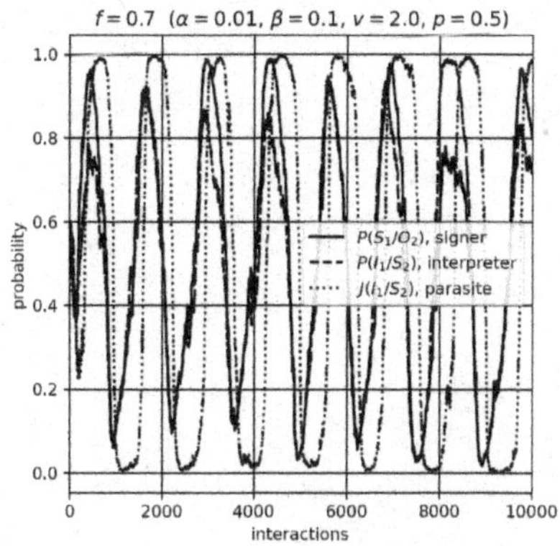

(b)

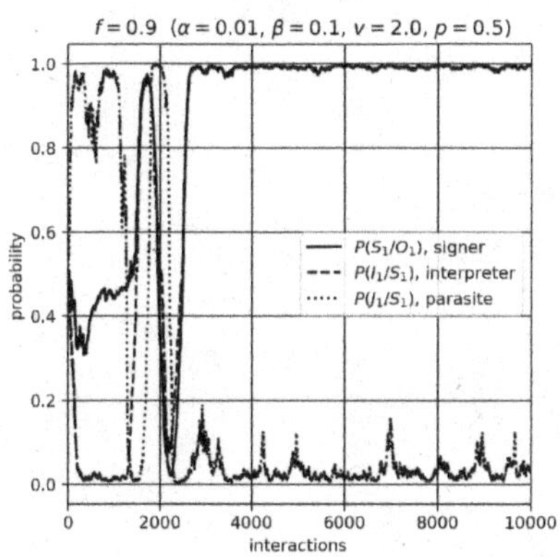

(c)

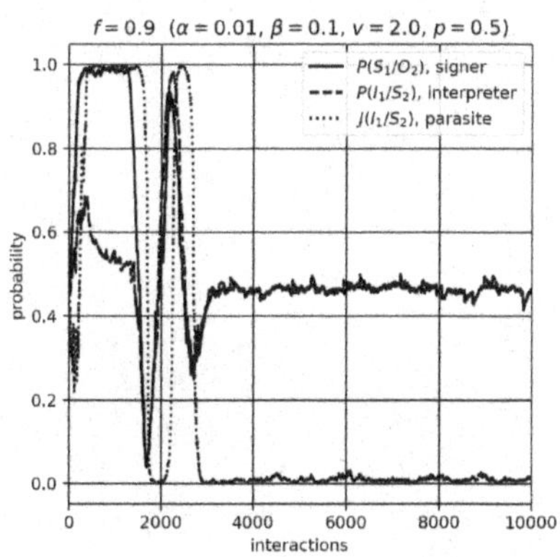

(d)

Figure 6.12: Semiotic Strategies, $f > .5$

7 Machine Semiosis

Overview

This chapter models semiotic agents as neural networks that are capable of learning from experience. In effect, the architecture of such networks, along with the values of the parameters they contain, constitute the interpretive grounds of these agents. Section 7.1 summarizes the core features of machine learning techniques as they are used in this chapter. Section 7.2 introduces readers to neural networks and their application to machine learning, building on the classic work of Rumelhart, Hinton, and Williams (1986). Using such networks, section 7.3 models a scenario in which two agents co-create an emergent code consisting of N arbitrary conventions. Section 7.4 calculates the average time that it takes to establish such a code from theoretical considerations, and compares it with the results of this scenario. Section 7.5 shows the relation between machine learning and meta-semiosis. And the conclusion discusses the relation between language models, which predict next words conditioned on prior words (independent of any world such words might refer to), and the mathematical model of meaning expounded in this book—in which worlds play a central role in the emergence of signs, minds, and social relations.

7.1 Machine Learning

Many forms of machine learning involve the following components. First, there is a set of data consisting of pairs of values, (x^i, y^i), where index i ranges over all pairs of values in the set ($i = 1, 2, \ldots, N$). These might be understood as the inputs (x) and outputs (y) of some as-yet-unknown function, which might represent, for example, the causes and effects of some force of interest.

Second, there is a function, $y_p = f(x; \theta)$, that is hypothesized to be a good model of the data. In what is known as *forward propagation*, such a function returns a predicted output, y_p, when given an input, x, and a list of parameters, θ. For example, having looked at the data, an analyst might assume that the relation between inputs and outputs is best modeled as a line, $y_p = mx + b$. In this relatively simple case, $\theta = \{m, b\}$, where m and b are two adjustable parameters, known as the slope and y-intercept, respectively.

Third, there is a function that measures the discrepancy between the predicted output and the actual output, given the hypothetical function, the data, and the current parameter values: $D(y_p, y) = D\big(f(x;\theta), y\big) = D(\theta)$. A simple measure of discrepancy, known as the *mean squared error*, is the average difference between y_p^i and y^i for all data points within the set:

$$D(\theta) = \frac{1}{N}\sum_{i=1}^{N}(y_p^i - y^i)^2 = \frac{1}{N}\sum_{i=1}^{N}\big(f(x^i;\theta) - y^i\big)^2. \tag{7.1}$$

Finally, there is an iterable procedure that updates the parameters of the function given the discrepancy:

$$\theta' = \theta - \alpha\nabla_\theta D(\theta). \tag{7.2}$$

In other words, the new parameter values, θ', are equal to the old parameter values, θ, minus the gradient of the discrepancy with respect to the parameters. Here, α is an adjustable meta-parameter known as the *learning rate*. The gradient, ∇_θ, is just the generalization of the derivative to multiple dimensions. When applied to the discrepancy, it returns a vector that represents how much the discrepancy changes as each of the parameters is changed. Phrased in another way, the gradient of the discrepancy is a vector that points in the direction of steepest increase (given the current values of the parameters). And so, by moving the values of the parameters in the opposite direction (hence the negative sign), the discrepancy is thereby lowered.

In short, with every iteration of this procedure, which is known as *back-propagation*, the value of each of the parameters is adjusted in a way that is designed to lower the discrepancy. And insofar as the discrepancy between predicted outputs and actual outputs is thereby lowered, the hypothesized function—with its updated parameters—becomes a better and better model of the data.

To apply machine learning techniques to semiotic processes, each of the components just discussed needs to be slightly transformed. The functions in question will be modeled as neural networks, and hence as relatively complicated non-linear functions that take vectors as inputs and return vectors as outputs. The parameters to be learned will be the weights and biases in such networks, and hence θ will be constituted by what is potentially a very long vector of adjustable values. Rather than start with already established data, an agent will generate data by interacting with the agents around it, as well as with the environment it inhabits, and then internalize the consequences of those actions (as a form of feedback). In other words, the agent does not need to know the correct interpretation of the signs that it receives in order to adjust its weights and biases; it only needs to know the value that it receives as a function of its interpretants. In effect, the agent will update, and thereby improve, its interpretive ground through a kind of reinforcement learning. Finally, rather than minimize discrepancy, the agent will maximize gain, which will be something like the expected value of its actions over time.

The next section offers a brief account of artificial neural networks, insofar as they are used in machine learning, and insofar as they constitute something like the physiology of such semiotic agents.

7.2 Neural Networks

Figure 7.1 shows a simple model of a single neuron as a mathematical function. As may be seen, the inputs correspond to dendrites, the function and parameters correspond to the nucleus, and the output corresponds to the axon.

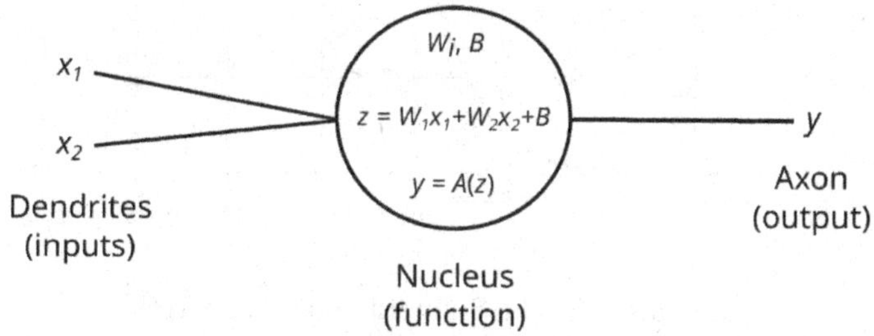

Figure 7.1: A Simple Model of a Single Neuron

Stated more carefully, every neuron in a neural network receives inputs, which may be represented as a vector of floating-point numbers: $x_1, x_2, \ldots, x_n$. Each of these inputs is multiplied by a neuron-specific weight $(W_1, W_2, \ldots, W_n)$; and a neuron-specific bias (B) is then added to the sum of those products, resulting in an intermediary value: $z = \sum_i W_i x_i + B$. That intermediary value becomes the argument of an activation function, $A(\cdot)$, whose output (itself another floating-point number, somewhere between zero and 1) constitutes the activation of the neuron: $y = A(z)$. For a given set of (positive) inputs, the greater the weights and/or the greater the bias, the greater this activation.

One commonly used activation function is the *sigmoid* function:

$$A(z) = \sigma(z) = \frac{1}{1+e^{-z}}. \tag{7.3}$$

As shown in figure 7.2, this function rapidly goes to zero as z goes to negative infinity; it rapidly goes to 1 as z goes to positive infinity; and it is linear, meaning $A(z) \approx z$, near zero. Such a function ensures that a neuron outputs a value near 1 when the sum of its weighted inputs (plus the bias) is positive; and it outputs a value near zero when the sum of its weighted inputs is negative. Its activation may thereby be framed as a probability distribution over a single outcome: true versus false (or do versus do not).

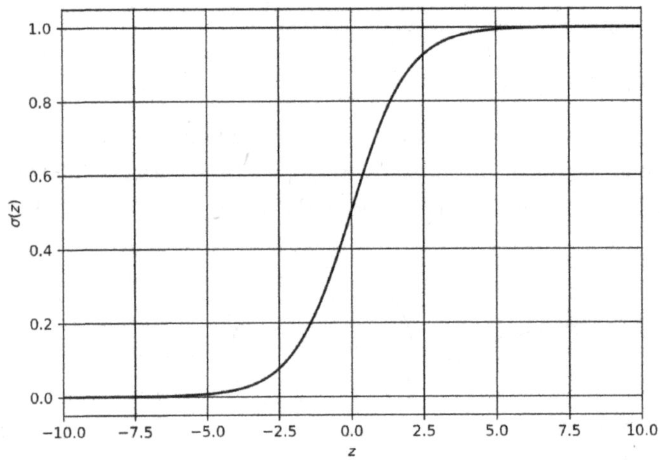

Figure 7.2: The Sigmoid Function, $\sigma(\cdot)$

Just as each input to a neuron can be the output of another neuron, the output of a neuron can constitute an input to other neurons (see figure 7.3). Such an arrangement of interconnected neurons is known as an *artificial neural network*. The rightmost layer of neurons in such a network is called the *output layer*. The next two layers would be called *hidden layers* (insofar as their outputs are available only to other neurons within the network). Finally, the leftmost layer is called the *input layer*. While conventionally (and somewhat confusingly) denoted with circles, the units in this leftmost layer are not actually neurons, but simply inputs to the network itself.

Whereas neurons in the hidden layers of a network often have the sigmoid function as their activation function, neurons in the final layer often have the softmax function as their activation function. As will be seen, this usage of the softmax function in the final layer allows us to treat the vector output of a neural network as a probability distribution over a set of outcomes.

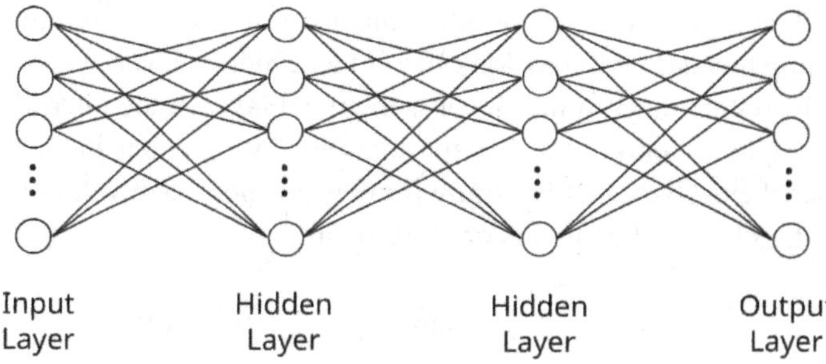

Figure 7.3: A Neural Network with Two Hidden Layers

Any such neural network can have any number of hidden layers, and any number of neurons in each of its layers (as indicated by the vertical dots in figure 7.3). Such an overall organization, or structure, is the architecture of the network. And each of the neurons in any network will have its own bias and weights. Depending on the values of all the weights and biases in a network, as well as its architecture, it will exhibit very different behaviors—understood as mappings from inputs to outputs.

In what follows, we will model semiotic agents as neural networks. Their inputs will be the signs that they sense. Their outputs will be the interpretants that they instigate. The weights and biases in such networks, in addition to the architecture of the networks themselves, will constitutes the interpretive grounds of such agents. Finally, if forward propagation (determining an output, given an input and already established parameter values) may be understood as a mode of semiosis, backpropagation (finding good parameter values for such a function) may be understood as a mode of meta-semiosis.

7.3 Establishing a Code

Let us revisit—and expand—a scenario that was introduced in chapter 1 and explored in chapter 6: two agents establishing a code consisting of N signs. Assume, in particular, that the signer can perceive objects and express signs and the interpreter can perceive signs and instigate interpretants. The two agents need to settle on a code to communicate about the location of a shareable resource.

As shown in figure 7.4, each agent will be modeled as a single-layer neural network. The signer takes in objects and puts out signs; the interpreter takes in signs and puts out interpretants. Here, O, S, and I are to be understood as lists of floating-point numbers (implicitly understood as probability distributions), where O_i denotes the value of the ith component of O (and it is similar for S and I). While this network may look like it has a hidden layer, it is simply the case that the output layer of the signer is the input layer of the interpreter. And while the diagram seems to localize the signs and interpretants in the bodies of the neurons themselves, that is simply for convenience; such values should be understood, rather, as the outputs of those bodies.

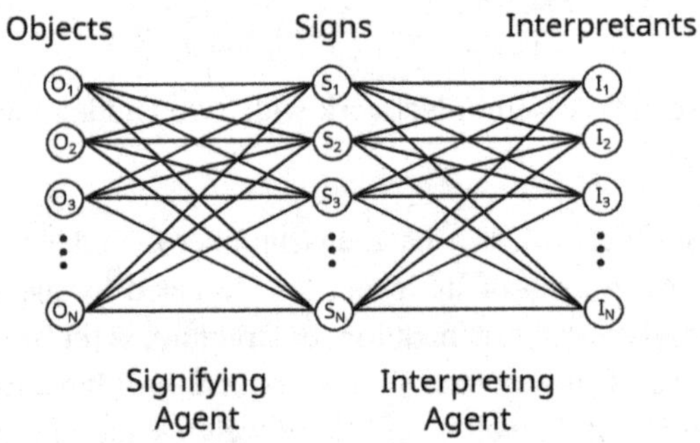

Figure 7.4: Signer and Interpreter as Coupled Neural Networks

Stated more carefully, the signifying agent, A^S, is constituted by the following function (understood as corresponding to the left-hand side of figure 7.4):

$$S_j = \text{softmax}(z_j^S) = \frac{e^{\beta z_j^S}}{\sum_k e^{\beta z_k^S}}, \tag{7.4}$$

where

$$z_j^S = \sum_l W_{jl}^S O_l + B_j^S, \tag{7.5}$$

and where O and S are vectors of length N, corresponding to the inputs and outputs of the network. Moreover, W^S is a matrix of size $N \times N$, and B^S is a vector of size N, corresponding to the learnable parameters in the network. As usual, β is an adjustable parameter of the softmax function.

Likewise, the interpreting agent, A^I, is constituted by the following function (understood as corresponding to the right-hand side of figure 7.4):

$$I_j = \text{softmax}(z_j^I) = \frac{e^{\beta z_j^I}}{\sum_k e^{\beta z_k^I}}, \tag{7.6}$$

where

$$z_j^I = \sum_l W_{jl}^I S_l + B_j^I. \tag{7.7}$$

Here, S and I are vectors of length N, corresponding to the inputs and outputs of the network. Moreover, W^I is a matrix of size $N \times N$, and B^I is a vector of size N, corresponding to the learnable parameters in the network.

Given the simplicity of the task to be learned, the bias terms, B^I and B^S, will be set to zero. This means that each agent has only N^2 parameters. In effect, each of the agents is really a function with an input, an output, and a relatively small matrix of learnable parameters:

$$S = A^S(O; W^S), \tag{7.8a}$$

$$I = A^I(S; W^I). \tag{7.8b}$$

As may be seen, the signer is essentially a function that takes in objects and turns out signs (given the current values of its parameters); and the interpreter is a function that takes in signs and turns out interpretants (given the current values of its parameters). Through their interactions, the two agents will learn relatively complementary parameter values that enable them to better secure and share the resource.

Before the agents begin to interact, and hence before they begin learning, the matrices W^S and W^I will be randomly initialized such that the parameters have an average value of 0 and a standard deviation of .01. This is somewhat equivalent to giving the agents a blank state.

Any given interaction between the two agents goes as follows: The resource is randomly assigned to one of N places with equal probability. O, as the input to A^S, is what is known as a *one-hot vector*: one of its values, which here corresponds to the location of the resource, is set to 1; and all its other values are set to zero. A lowercase o will be used to denote the index corresponding to the object's location in a given interaction, such that $O_j = \delta_{oj}$. (As usual, δ_{ij} is the Kronecker delta function, which is 1 when $j = i$ and zero otherwise.)

S, as the output of A^S, is thus a probability distribution or set of 'activations' over possible signs that could be expressed (given the object). It will be assumed that the signer always acts on the most probable of these signs. In other words, the actual sign expressed by the signer (as opposed to its outputted probability distribution), denoted S^{act}, which goes on to become the input to the interpreter, is also a one-hot vector. A lowercase s will be used to denote the index of the actualized sign in a given interaction, such that $S_j^{\text{act}} = \delta_{sj}$.

Similarly, I, as the output of A^I, is a probability distribution over possible interpretants that could be instigated (given the sign). It will be assumed that the interpreter always acts on the most probable of these interpretants, and hence the most activated of its neurons. In other words, the actual interpretant instigated by the interpreter (as opposed to its outputted probability distribution), denoted I^{act}, which goes on to determine the payoff for both agents, is also a one-hot vector. A lowercase i will be used to denote the index of the actualized interpretant in a given interaction, such that $I_j^{\text{act}} = \delta_{ij}$.

Finally, depending on the relation between the location of the resource and the action of the interpreter, both agents may receive a relatively generous reward (half the value of the resource, v) and incur a small cost ($c < v$). In particular, it is assumed that each agent receives the following reward as a kind of feedback:

$$V_{io} = \frac{v}{2}\delta_{io} - c \,. \tag{7.9}$$

In other words, both agents gain a value of $v/2$ if they successfully communicate (such that the interpreter goes to the location of the resource, meaning $i = o$); and both incur a small cost (as a result of their communicative efforts, understood as the labor of signifying and interpreting), regardless of the outcome.

It should be emphasized that neither agent knows which output it *should* have generated (understood as the correct sign or interpretant). All an agent has access to are its own inputs and outputs, its weights and architecture (and hence its own functional specification, as embodied in its neural network), and what reward, or cost, it ultimately receives or incurs (V_{io}). We may therefore define a gain function for each agent as follows:

$$G^S = V_{io}\, S_s \,, \tag{7.10a}$$

$$G^I = V_{io}\, I_i \,. \tag{7.10b}$$

In other words, the gain to the signer, G^S, is simply the value that goes to the signer at the end of an interaction (which itself is a function of the relation between the object's position and the interpreter's action), multiplied by the probability, or activation level, of the sign that it expressed during the interaction. And it is similar for the gain to the interpreter. (It should be stressed that these gain functions turn on S and I, not S^{act} and I^{act}.)

While these functions may look mysterious, they are easily motivated. If conditional probabilities were being used, as was done in chapter 6, the expected value for each agent (during a given interaction) would be as follows:

$$\langle V \rangle = \sum_{jkl} V(I_l \wedge O_j) \cdot P(O_j) \cdot P(S_k/O_j) \cdot P(I_l/S_k)\,. \tag{7.11}$$

Translating this equation into the terms of the two agents, understood as neural networks, as well as the payoff matrix introduced here, we get

$$\langle V \rangle = \sum_{jkl} V_{lj} \cdot O_j \cdot S_k(O) \cdot I_l(S)\,. \tag{7.12}$$

As may be seen, we are emphasizing the fact that S is a function of O and I is a function of S. Recall equations (7.4) and (7.6). During a given interaction, O is a one-hot vector, and S and I become one-hot vectors (insofar as each agent acts on the most activated neuron in its final layer), so each of the terms in the sum is zero except for those that are actualized during a given interaction:

$$\langle V \rangle = V_{io} \cdot O_o \cdot S_s(O) \cdot I_i(S)\,. \tag{7.13}$$

From the standpoint of the signer, $O_o = 1$ and $I_i(S) \rightarrow I_i^{\text{act}} = 1$, which gives G^S. And from the standpoint of the interpreter, $O_o = 1$ and $S_s(O) \rightarrow S_s^{\text{act}} = 1$, which gives G^I. In short, this measure of gain is just a variant of expected value, as motivated by the main equation from chapter 1.

Following the discussion of gradients and the updating of parameter values, it will be assumed that the change in an agent's weights is proportional to the gradient of the gain with respect to the weights: $\nabla_{W^I} G^I$. Thus, after each interaction, the weights in the interpreter's neural network are adjusted in the following manner:

$$\begin{aligned} W'^{I}_{pq} &= W^I_{pq} + \alpha \nabla_{W^I_{pq}} G^I \\ &= W^I_{pq} + \alpha V_{io}\, \nabla_{W^I_{pq}} I_i \\ &= W^I_{pq} + \alpha V_{io}\, \frac{\partial I_i}{\partial z^I_p} \frac{\partial z^I_p}{\partial W^I_{pq}} \\ &= W^I_{pq} + \alpha V_{io}\, I_i\,(\delta_{ip} - I_p)\, S^{\text{act}}_q \\ &= W^I_{pq} + \alpha V_{io}\, I_i\,(\delta_{ip} - I_p)\, \delta_{sq}\,. \end{aligned} \tag{7.14}$$

As may be seen in the second line in equation (7.14), we are taking for granted the fact that G^I depends on the parameters insofar as $I_i = I_i(S; W^I)$. The chain rule from calculus is used in the third line. The derivative of the softmax function is used in the fourth line. Finally, the fifth line takes into account the fact that the sign received by the interpreter is a one-hot vector: all its values are zero except for the value that corresponds to the sign that the signer expressed during an interaction.

Rewriting the last line, and letting $\Delta W^I_{ps} = W'^I_{ps} - W^I_{ps}$, results in the following key equation:

$$\Delta W^I_{ps} = \alpha V_{io} I_i (\delta_{ip} - I_p) \,. \tag{7.15}$$

In other words, after a given interaction, only the weights in the interpreter's network that correspond to the expressed sign are altered (and hence only a single column in the interpreter's matrix of weights is affected, corresponding to index *s*). And for each entry in that column, corresponding to a different interpretant (as indexed by *p*), the weights are increased if the value received was positive, and the interpretant is the one that was instigated ($p = i$), and lowered otherwise. If the value received was negative, in contrast, the weights corresponding to a given interpretant are reduced if that interpretant was the one instigated, and increased otherwise. Finally, all this is done in proportion to the activation levels, or probabilities, of the neurons in question. This equation constitutes the learning rule for the interpreting agent.

An almost identical derivation leads to the update rule for the signer's weights:

$$\Delta W^S_{po} = \alpha V_{io} S_s (\delta_{sp} - S_p) \,. \tag{7.16}$$

In other words, after a given interaction, only the weights in the signer's network that correspond to the actual position of the object are altered (and hence only a single column in the signer's matrix of weights is affected, corresponding to index *o*). And for each entry in that column, corresponding to a different sign, the weights are increased if the value received was positive and the sign is the one that was expressed, and reduced otherwise. (And conversely, as discussed previously, if the value received was negative.) This equation constitutes the learning rule for the signifying agent.

Equations (7.15) and (7.16) should be compared with the procedure that was used for updating the weights in an agent's register in chapter 6, when traditional forms of reinforcement learning were explored. In particular, there the value that was received, be it positive or negative, would simply be added to the appropriate weight in the agent's register. Here, in contrast, all the weights in a column are adjusted as a function not just of the value received, but also by the current probabilities of possible actions. As will be seen, this is a more efficient—and general—way for an agent to learn.

The signifying and interpreting agents now have all they need to establish a code consisting of N signs by means of interacting with each other and embodying, in their neurons, the rewards that they receive. We now want to see how many interactions it takes, on average, for a pair of such agents to establish such a code.

Rather than track the expected value, or gain, of the agents as a function of interaction number, we will track the percentage of the code they successfully establish. In particular, after each interaction, as a measure of how well they are learning, we put the object in each of the N possible locations; we let the signer encode its location (using its current weights); we let the interpreter decode that sign (using its current weights); and we ask whether the interpreter ended up securing the resource by going to the right location. If $N = 10$, for example, and the interpreter went to the right location for 8 of the 10 possible positions of the object, then we will say that the agents have successfully established 80 percent of the code. In other words, 8 of the 10 signs led the interpreter to the right place, such that the interpreter's mapping from S to I was the inverse of the signer's mapping from O to S.

Figure 7.5 plots such percentages as a function of the number of interactions that a pair of agents has had with each other. The solid black lines show the average percentage of a code that was successfully established, per interaction, by a population of 200 pairs of agents (where each pair, consisting of a signer and an interpreter, is learning independently from the others). The thinnest line shows pairs of agents establishing codes with 4 symbols; the thickest line shows pairs of agents establishing codes with 16 symbols; and it is similar for the lines with intermediate thickness. As may be seen, and as will be discussed at length in section 7.4, it takes progressively longer for agents to establish larger codes. The three dotted black lines show the percentage of the code that was successfully established by individual pairs of agents (as opposed to a population of 200 pairs of agents), establishing codes with 16 symbols. As may be seen, there is a wide variation from the average (as plotted by the thick, solid line): two pairs of agents learned much faster than average and one pair of agents learned much, much slower.

Moving from left to right, the black dots at the top of figure 7.4 show the expected time-to-code, $T_c(N)$, for a pair of agents establishing a code with N arbitrary symbols (when $N = 4, 6, \ldots, 16$), where

$$T_c(N) = N^3 \sum_{n=1}^{N} \frac{1}{n^2}. \tag{7.17}$$

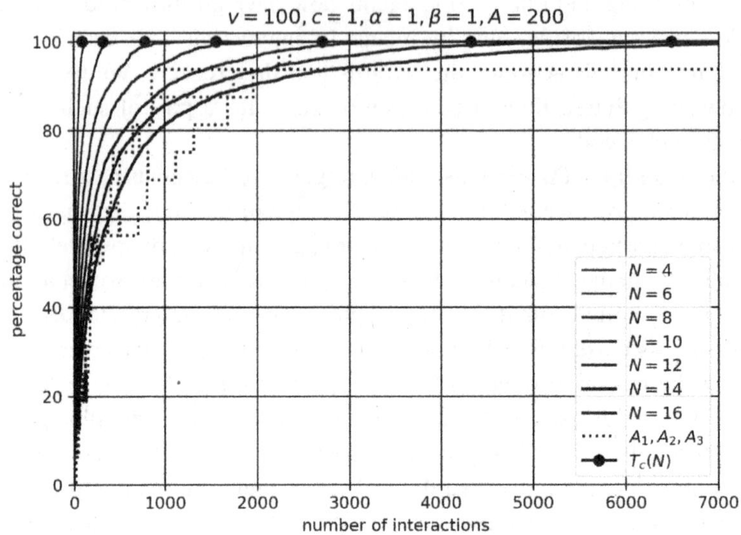

Figure 7.5: Signer and Interpreter Establishing a Code with N Symbols

As may be seen, assuming that the value is very high ($v = 100$) and the cost is relatively low ($c = 1$), the pairs of agents—on average—establish a shared code near this predicted rate.

The next section will derive equation 7.17, and motivate its application to this scenario.

7.4 The Time It Takes to Establish a Code

Within the bounds of the foregoing scenario, when signifying and interpreting agents establish a code, they are effectively determining a mapping from objects to signs, along with an inverse mapping from signs to interpretants. Expressed functionally, $I(S(O)) = O$, such that $I = S^{-1}$ (so to speak).

To make this last point clearer, here are the weights—and hence final parameter values—in the networks of a signer and interpreter (rounded to the nearest integer) after they had learned a code with four signs:

$$W^S = \begin{bmatrix} -7 & -8 & -7 & 21 \\ -7 & -6 & 20 & -7 \\ 21 & -7 & -6 & -7 \\ -7 & 20 & -7 & -7 \end{bmatrix}, \tag{7.18a}$$

$$W^I = \begin{bmatrix} -7 & -6 & 20 & -6 \\ -6 & -7 & -7 & 20 \\ -7 & 20 & -6 & -7 \\ 19 & -7 & -7 & -7 \end{bmatrix}. \tag{7.18b}$$

And here are the softmax functions of those weights, corresponding to $S(O)$ and $I(S)$:

$$S = \text{softmax}(W^S) \approx \begin{bmatrix} 0 & 0 & 0 & 1 \\ 0 & 0 & 1 & 0 \\ 1 & 0 & 0 & 0 \\ 0 & 1 & 0 & 0 \end{bmatrix}, \tag{7.19a}$$

$$I = \text{softmax}(W^I) \approx \begin{bmatrix} 0 & 0 & 1 & 0 \\ 0 & 0 & 0 & 1 \\ 0 & 1 & 0 & 0 \\ 1 & 0 & 0 & 0 \end{bmatrix}. \tag{7.19b}$$

As may be seen from the matrix entries in equation (7.19), the signer ended up mapping object 1 onto sign 3, object 2 onto sign 4, object 3 onto sign 2, and object 4 onto sign 1; and the interpreter ended up doing the inverse of those mappings. In other words, and as would be expected, if the sign indicates the location of the food, the interpretant is the action of going to that location.

It should be emphasized that this is only one of $4! = 24$ possible codes that the two agents could have settled on. More generally, supposing a scenario that involves N signs (and hence N objects and N interpretants), there are $N!$ possible mappings that a signer and interpreter might eventually settle on. This is because the first object can be mapped onto any of N signs; the second object can then be mapped onto any of $N-1$ signs; the third object can then be mapped onto any of $N-2$ signs; and so forth. In effect, the signer has $N \cdot (N-1) \cdot (N-2) \cdot \ldots \cdot 1$ possible maps at their disposal. And, for each of these $N!$ possible encoding schemes, there is one (and only one) possible decoding scheme that may be employed by the interpreter (the inverse of that encoding scheme), insofar as the gain has been maximized.

To be sure, it is not the case that the signer first determines a mapping, which the interpreter then proceeds to learn. Rather, the agents establish a shared mapping—over interactional time—by updating the weights in their neural networks, as a function of the success of their past interactions (as measured by their respective gains), according to the method derived in section 7.3. Thus, there is a lot of trial and error.

We now want to estimate the number of interactions its takes—on average—for the agents to settle on one of these $N!$ possible codes. In effect, we want to determine the time it takes to establish a language with N signs.

To calculate this average time-to-code, as a function of N, which was denoted as $T_c(N)$, we begin with two assumptions. The first is well known from probability theory: when rolling an unbiased N-sided die, it takes N rolls on average for a given number to come up. More generally, if something has a probability p of happening during a given unit of time, and happenings during different times are uncorrelated, then it takes $1/p$ units of time on average for it to happen. The second assumption is that the agents need only one successful communicative encounter to establish a sign's encoding; and once established, they never waver from it again.

Suppose, for example, that the signer happens to produce sign 3 when it perceives that the object is located at position 2, and the interpreter happens to go to position 2 when it senses sign 3. Both agents will thereby benefit from the communicative encounter, and thus the weights in their networks will change accordingly. Our second assumption is essentially that the weights change in such a way that, ever after, $S_3(O)$ is max when $O = \delta_{2j}$ and $I_2(S)$ is max when $S = \delta_{3j}$. Phrased in another way, W^S_{32} has the largest value in its column and W^I_{23} has the largest value in its column. Given the fact that each agent acts on its most activated neuron, meaning the most probable S for the signer and the most probable I for the interpreter, they will end up repeating the successful behavior in later interactions.

To calculate $T_c(N)$, assume that the weights in the agents' networks are initialized randomly, so the agents are equally likely to engage in any possible behavior. In any given communicative encounter, there are N^3 possible outcomes, or worlds: one of N objects gives rise to one of N signs, which gives rise to one of N interpretants. Crucially, N of these encounters are successful, insofar as the interpreter goes to the right location, such that both agents benefit, such that the weights in their neural networks—and the attendant behaviors—change accordingly. In short, before the agents have established the meaning of any signs, the probability that the encounter is successful is $N^2/N^3 = 1/N$, so it takes N encounters, on average, for the interpreter to end up going to the right location, whatever the sign happens to be. And hence it takes N encounters, on average, to establish the first shared convention.

Once the agents have established the object of one sign, they need to establish the object of another sign out of the $N-1$ remaining signs. The chance that they have a successful encounter with a sign whose object has not yet been established is thus $(N-1)^2/N^3$, and hence lower than before. So it now takes $N^3/(N-1)^2$ encounters, on average, to establish the object of a second sign.

To establish the object of a third sign will require $N^3/(N-2)^2$ encounters, on average, and so forth. As may be seen, it takes longer and longer to establish the object of the next sign. And to establish a mapping—and hence a designated object—for all the signs in a language with N signs requires, on average, the following number of communicative encounters:

$$T_c(N) = N^3 \sum_{n=1}^{N} \frac{1}{n^2} = N^3 \left(\frac{1}{1^2} + \frac{1}{2^2} + \ldots + \frac{1}{N^2} \right). \tag{7.20}$$

In the limit of large N, the series converges to $\pi^2/6$, such that

$$T_c(N) \approx \frac{\pi^2}{6} N^3 \text{ for large } N. \tag{7.21}$$

In short, as N increases, not only does it take an increasingly long time for a pair of agents to create a shared code, and thereby establish a set of conventions (going as N^3), it also takes a pair of agents progressively longer to establish the next convention in their set (whatever the value of N), insofar as the agents are less and less likely to have a novel interaction, or outcome, that they can learn from. This is why the learning curves flatten as time goes on.

As was seen in figure 7.5, the average time-to-code for an ensemble of 200 agents (the black lines) was in surprisingly good agreement with these estimates. That said, it would not be wise to put too much weight on the precise correspondence, which may be sensitive to the value of variables like c and v. More important is that the calculated estimate captures the correct ratios, and the fact that time-to-code goes as N^3.

It should be emphasized that this method of establishing a shared code, although much better than traditional forms of reinforcement learning, is not particularly efficient. Note, for example, that even though the agents may have settled on a particular mapping (say, object 1 maps to sign 3 and sign 3 maps to interpretant 1), the signer will still try sign 3 in the context of other objects, and the interpreter will still try interpretant 1 in the context of other signs. In other words, while the agents generalize across columns (in their respective weights, W^S and W^I), they do not generalize across rows. If they could generalize across rows, then they would learn much faster in this kind of scenario, for the signer would never use an already established sign for a different object and the interpreter would never use an already established interpretant for a different sign. Such options would be ruled out. So the agents would more quickly generate possible outcomes that they could learn from, and thereby more quickly establish a shared code.

To be sure, there are many other reasons why this is not the most plausible scenario as it now stands. For example, it presumes that the resource is equally likely to be found in any location; and it presumes that the signs are completely arbitrary. As discussed in chapter 6, certain objects are often more likely than others in a given environment, and certain signs are often more motivated (for a given object) than others. And agents can use such correlations to their advantage and thereby establish shared codes faster. Given that the environments of most semiotic agents have such biases and costs built into them, we would expect that the emergence of semiotic codes would reflect such patterns. Finally, we are modeling our semiotic agents as extremely simple (and shallow) neural networks. It is relatively easy to introduce agents with more complicated architectures, and larger numbers of adjustable parameters. In any case, the foregoing scenario, along with its agents, may be readily adjusted to handle such additions, while still retaining the benefits of semiotic agents that are artificial neural networks, by taking the lead from the more complicated scenarios analyzed in chapter 6.

The next section will reframe forward propagation as a mode of semiosis and backpropagation as a mode of meta-semiosis.

7.5 Machine Learning as Meta-Semiosis

This chapter has modeled semiotic agents as neural networks and used techniques from machine learning to train them to signify and interpret. We watched as two agents created an emergent set of conventions, or symbolic code, to coordinate their actions to acquire a resource. In so doing, we recast the learned parameter values in a network, along with the architecture of the network itself, as the interpretive ground of the agent that embodied the network. We also derived a simple yet powerful algorithm for updating parameter values in the context of value-mediated feedback. Before continuing, it is worth drawing out a few more connections among machine learning, neural networks, and semiotic processes.

As was shown, an (interpreting) agent engaging in machinic semiosis may be understood as a parameter-dependent function that takes signs as inputs and returns interpretants as outputs: $I = A(S; \theta)$. Stated more carefully, we examined scenarios in which: signs were caused by, and/or correlated with, certain objects; and the interpretants instigated by the agents, as governed by the parameter values that they learned, came to make sense in the context of the signs that they sensed, given the relation between the values of the objects and the interests of those agents. In effect, $A(\cdot)$ was not just an immediate function of S; it was also a more mediate function of O and V.

Crucially, it is not just the case that semiotic agents can be modeled as neural networks. It is also the case that machine learning is a mode of semiosis per se. To see how, let us return to the example from section 7.1, which turned on a set of data, (x^i, y^i); a parameterized function, $y_p = f(x; \theta)$; a measure of discrepancy, $D(\theta)$; and an algorithm for updating the parameters to reduce the discrepancy, $\Delta\theta = -\alpha\nabla_\theta D(\theta)$. In such a situation, the sign may be identified with the input, x^i. The actual correlate, y^i, may be identified with the object. And the predicted correlate, $y_p^i = f(x^i; \theta)$, may be identified with the interpretant. In other words, forward propagation—whereby an interpretant (or output) is calculated given a sign (or input)—is itself a semiotic process (see figure 7.6).

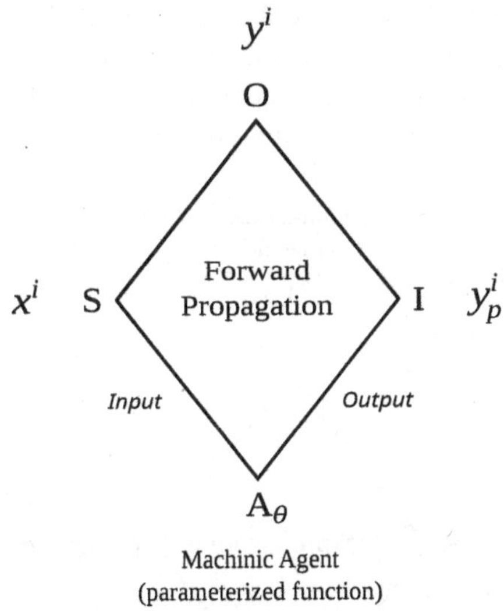

Figure 7.6: Forward Propagation as a Semiotic Process

All this should make sense. Any process that goes beyond the given, and hence engages in some mode of induction (from part to whole, from sample to population, from past observations to future predictions, and from training data to test data), is inherently semiotic, insofar as it treats the given as evidence for what is conveyed by, but not contained in, the given.

Even more crucially, perhaps, is the fact that backpropagation is a mode of meta-semiosis. To see how, first note that the derivative of a function at a specific point, $f'(a)$, provides information about the function in the vicinity of the point. In particular, by knowing the value of the function at a point and the derivative of the function at that point, we can estimate the value of the function near that point: $f(a+\Delta x)\approx f(a)+f'(a)\Delta x$. In effect, the derivative of a function at a point constitutes a sign, the actual value of the function near that point constitutes the object, and the estimated value of the function near that point—or some action that turns on such an estimate—constitutes the interpretant.

When we take the gradient of the discrepancy at a given set of parameter values, for example, we are essentially finding the slope of the discrepancy at a certain point in parameter space. Here, then, $\nabla_\theta D(\theta)$ is the sign. The shape of the discrepancy function near θ is the object. And the change in parameter values, as an action designed to lower the discrepancy, is the interpretant.

Indeed, as shown in figure 7.7, when this process is repeated again and again, the agent essentially finds its way to the lowest point—or at least a local minimum—in parameter space. Such a point not only constitutes an object, it also constitutes an objective: the place where the meta-agent is trying to go, so to speak, by repeatedly moving in the most downward-pointing direction.

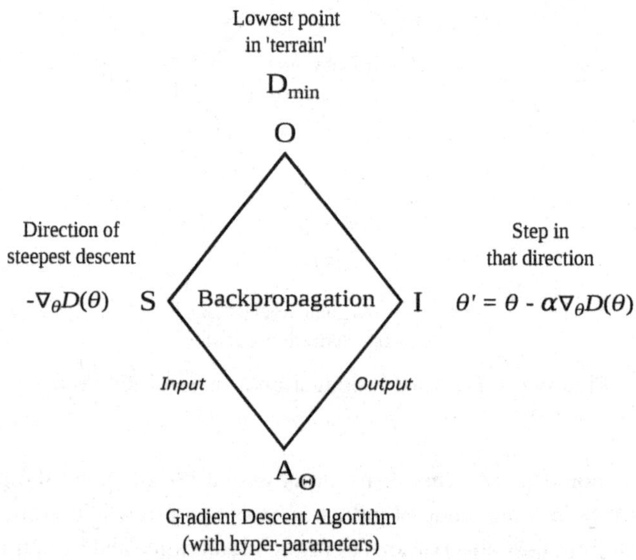

Figure 7.7: Backpropagation as a Meta-Semiotic Process

Such a process is analogous to a mapless hiker using the shape of their immediate surroundings to find the nearest valley: Find the direction of steepest descent where they currently are, take a step in that direction, and repeat. In other words, local clues lead to a relatively remote and desirable destination.

7.6 Language Models

Given all the media attention to ChatGPT and other recent developments in artificial intelligence, it is worth saying a few words about language models. Kockelman (2024) analyzes their inner workings and social implications in depth. So the following discussion is confined to a single dimension of contrast: they model language with minimal reference to the world. In this way, language models are decidely different from the mathematical models that have been the focus of this book, which model meaningful processes in relation to the patterns and resources found in worlds, as well as in relation to the social relations of the agents that both reside in and represent such worlds.

(Moreover, whereas language models now have billions, if not trillions, of parameters, we have endeavored throughout this book to model meaningful processes with as few parameters as possible.)

Stated more carefully, while our model of meaning has focused on relations between signs and objects (as well as relations between interpretants and consequents), and hence on the coupling between representations and worlds, language models focus on word-word relations. In particular, the core capability of such models is next-word prediction.

Framing their capacity in terms of our model of meaning, the sign is a sequence of words from a human-authored text: $w_1, w_2, \ldots, w_n$. The object is the actual next word in the sequence (as it occurs in the text): w_{n+1}. And the interpretant is a probability distribution over possible next words (conditioned on the preceding words in the sequence): $P(w_{n+1}|w_1, w_2, \ldots, w_n)$. The key capacity of such agents is thus easily captured in our model's terms (see figure 7.8).

Once a model has been trained on a corpus of human-authored texts, by adjusting its parameters until it can accurately undertake such predictions (using backpropagation to minimize a discrepancy function known as cross-entropy loss), it can then be used to generate text, and thereby partake in certain kinds of pseudo conversations. In effect, it generates next moves conditioned on prior moves, and thereby produces uncannily real interactions, by conforming with the textual patterns embodied in that corpus.

The intentionality of such models—the way that their signs seem to be directed at objects, or the way that their behavior seems to be oriented towards ends—is thus wholly parasitic on human intentionality. Framed in another

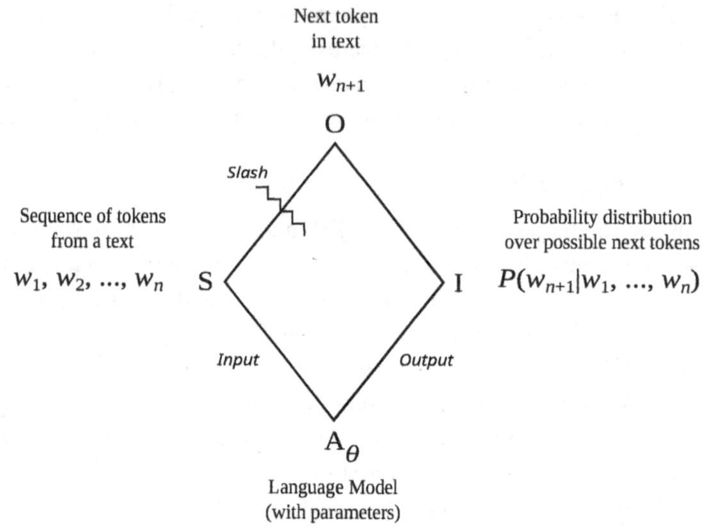

Figure 7.8: Next-Word Prediction as a Semiotic Process

way, the interpretive grounds of such models are parasitic on the interpretive grounds of the humans who trained the models, wrote the texts that the models were trained on, or interact with the models through prompts and responses.

As should be clear from figure 7.8, the slash separating the sign and the object in language models, and hence that which separates what is given from what must be inferred, is not the kind of slash that separates representations from the world or actions from intentions; rather, it is the kind of slash that separates earlier parts of texts from later parts, and hence past words from future words.

Language models are thus quite distinct from the kinds of semiotic agents that have interested us in this book—agents whose actions and inferences are tightly coupled to worlds (and worldviews in the guise of interpretive grounds). In particular, indexical reference (to the world) and true representations (of the world) are beyond such models. In a certain sense, then, the key capacity of language models (next-word prediction) is also their main limitation: *worldlessness*. Our mathematical model of meaning, in contrast, has put the world—which includes existing (and emergent) social relations, as well as worldviews—at the center of meaning.

This concludes our discussion of agents that learn. We now turn to some of the presuppositions and extensions of this model of meaning.

IV PRESUPPOSITIONS AND EXTENSIONS

8 Possible World Semiotics

Overview

This chapter is about possible worlds and their relation to meaningful practices. Such worlds might be provisionally understood as states that a system can show up in, or circumstances of evaluation for the truth of a proposition. Section 8.1 introduces the concept of possible worlds using standard conventions. Section 8.2 briefly connects possible worlds to formal semantics, with a particular focus on modal operators. Section 8.3 projects probabilities onto possible worlds so the analysis can be done quantitatively. Section 8.4 relates possible worlds to the mathematical model of meaning introduced in chapter 1, with a particular focus on Bayesian networks (or causal graphs) and expected value. Section 8.5 reviews Angelika Kratzer's notion of a conversational background, as well as her notions of possibility and necessity, and shows the relation between such concepts and interpretive grounds more generally. Section 8.6 analyzes the relation between times and worlds, as a means to track the changing interpretive grounds of agents as they come to know and transform their environments. Finally, section 8.7 builds on these ideas to show the relation between the changing grounds of agents and the changing models of analysts.

8.1 Introduction to Possible Worlds

To see the value of possible worlds as an analytic device, it is helpful to start with a concrete example. Recall the opening scenario, from chapter 2, which turned on the following ontology:

O = {person is healthy, person is sick}

S = {person does not evince symptom, person evinces symptom}

I = {person does not take medication, person takes medication}

$C = \{$person remains (or becomes) healthy, person remains (or becomes) sick$\}$

With such a scenario in mind, a few preliminary terms may be defined. As may be seen, there are four *variables* in this ontology: O, S, I, and C. If the *domain* of a variable X, denoted as dom(X), is the set of values that the variable can take, these variables have the following domains: dom(O) = $\{O_1, O_2\}$, dom(S) = $\{S_1, S_2\}$, dom(I) = $\{I_1, I_2\}$, and dom(S) = $\{C_1, C_2\}$. As may also be seen, to avoid having to spell out the propositional contents of such values (e.g., "person is healthy"), they will usually be referred to using the variable name, along with a subscripted index (e.g., O_1).

A *possible world* assigns one value to each such variable, given such a scenario and its attendant ontology. To specify each of the possible worlds in this opening scenario, the following terminology will be used: Let w_{ijkl} be the world in which C takes the value C_i, I takes the value I_j, S takes the value S_k, and O takes the value O_l. For example, w_{2121} is the world in which C is assigned the value C_2, I is assigned the value I_1, S is assigned the value S_2, and O is assigned the value O_1. In other words, it is the world in which the person is sick, the person does not evince the symptom, the person does not take the medication, and the person remains sick.

Given the domains of these variables, each of the four indices (i, j, k, l) can only take one of two values (1 or 2). This results in a scenario-specific *universe*, which will be called Ω (omega), with the following possible worlds:

$$\begin{aligned}\Omega = \{&w_{1111}, w_{1112}, w_{1121}, w_{1122},\\ &w_{1211}, w_{1212}, w_{1121}, w_{1222},\\ &w_{2111}, w_{2112}, w_{2121}, w_{2122},\\ &w_{2211}, w_{2212}, w_{2121}, w_{2222}\}.\end{aligned} \tag{8.1}$$

As may be seen, Ω is a complete set of possible worlds, and hence a universe, whose numerical size or *cardinality*, denoted as $|\Omega|$, is $2^4 = 16$. As may also be seen, notwithstanding their seemingly mysterious and metaphysical nature, possible worlds are simply states that a system can show up in (within a particular framing).

When describing possible worlds, the expression $X(w) = x$ will be used to mean that variable X is assigned value x in world w. Given the foregoing assignments, for example, it can be said that

$$C(w_{ijkl}) = C_i\,, \tag{8.2a}$$

$$I(w_{ijkl}) = I_j\,, \tag{8.2b}$$

$$S(w_{ijkl}) = S_k\,, \tag{8.2c}$$

$$O(w_{ijkl}) = O_l\,. \tag{8.2d}$$

Such a convention allows sets of worlds to be specified in relatively simple ways. Take, for example, the following expression:

$$\{w \mid (C(w) = C_1) \wedge (I(w) = I_2)\} = \{w_{1211}, w_{1212}, w_{1221}, w_{1222}\}. \tag{8.3}$$

Here, the expression on the left of the equality should be read as saying 'the set of worlds w in which $C(w)$ takes the value C_1 and $I(w)$ takes the value I_2'; and the expression on the right of the equation is just the subset of worlds from the universe that actually fulfills these conditions. Loosely, if the left-hand side of the equation defines the set intensionally, the right-hand side delimits the set extensionally.

Having described worlds as assignments of values to variables, a *proposition* may be understood as a set of worlds. For example, proposition O_1 is the set of worlds in which O_1 obtains, and hence the set of worlds in which "person is healthy" is true. More specifically, and within the universe Ω of 16 possible worlds, this means

$$\begin{aligned} O_1 &= \{w \mid O(w) = O_1\} \\ &= \{w_{1111}, w_{1121}, w_{1211}, w_{1221}, w_{2111}, w_{2121}, w_{2211}, w_{2221}\}. \end{aligned} \tag{8.4}$$

Similarly, proposition $I_1 \wedge S_2 \wedge O_1$ is the set of worlds in which I_1, S_2, and O_1 obtain. More specifically, this means

$$\begin{aligned} I_1 \wedge S_2 \wedge O_1 &= \{w \mid (I(w) = I_1) \wedge (S(w) = S_2) \wedge (O(w) = O_1)\} \\ &= \{w_{1121}, w_{2121}\}. \end{aligned} \tag{8.5}$$

In other words, $I_1 \wedge S_2 \wedge O_1$ is the set of worlds in which it is true that the person is healthy, the person evinces the symptom, and the person does not take the medicine.

The *power set* of a set S, denoted as $\mathcal{P}(S)$, is the set of all subsets of that set, which includes the empty set and the original set itself. Insofar as Ω is a complete set of possible worlds, $\mathcal{P}(\Omega)$ is the set of all sets of possible worlds, and hence the set of all possible propositions. As is well known, the cardinality of a set relates to the cardinality of its power set by the following relation:

$$\left|\mathcal{P}(S)\right| = 2^{|S|}. \tag{8.6}$$

Therefore, given the fact that $|\Omega| = 16$, the cardinality of its power set is

$$\left|\mathcal{P}(\Omega)\right| = 2^{|\Omega|} = 2^{16} = 65,536. \tag{8.7}$$

Thus, even in the context of a relatively tiny universe, it is easy to conceive of a relatively large number of possible propositions, where any such proposition may be true or false depending on the actual state of the system.

In short, a possible world is not just a possible state or outcome of a system (as well as a complete description of all the facts that obtain, under a certain framing). It may also be understood as a circumstance of evaluation for the truth of a proposition (as well as for the utility, and/or felicity, of an action).

To clarify this last point, the next section turns from the logical specification of possible worlds to the role that such worlds play in a certain theory of semantic meaning.

8.2 Possible World Semantics

Shifting for a moment to possible world semantics, a proposition may be understood as a function from worlds to truth values. In other words, proposition p accepts world w as an argument and returns the value 1 ("true") if the world is in the set delimited by the proposition, and the value 0 ("false") if the world is not in that set. The returned value is typically called the *denotation* of the proposition. More formally, if α is a sentence, uttered in world w, that expresses proposition p, it may be said that

$$[\![\alpha]\!]^w = \begin{cases} 1 & w \in p, \\ 0 & w \notin p. \end{cases} \tag{8.8}$$

Here, the doubled brackets $[\![\cdot]\!]$ indicate the operation of denotation, such that this formalism should be read as follows: "the denotation of sentence α, as uttered in world w, is true if w is in p, and false otherwise" (where p is to be understood as the proposition expressed by α, and hence as a set of possible worlds that may—or may not—contain w).

Suppose, for example, that α is a sentence (in some language) that expresses proposition O_1. And suppose that this sentence is uttered in world w_{1111}, which is thereby understood as something like the speech event, circumstance of evaluation, or world of narration, for the utterance in question. Given the way that O_1 and w_{1111} were defined in the preceding section (such that $w_{1111} \in O_1$), we find that

$$[\![\alpha]\!]^{w_{1111}} = 1\,. \tag{8.9}$$

If, in contrast, α expressed proposition $I_1 \wedge S_2 \wedge O_2$, and it was uttered in the same world, then we would have found that

$$[\![\alpha]\!]^{w_{1111}} = 0\,. \tag{8.10}$$

In other words, one and the same proposition may be true when evaluated in one world, but false when evaluated in another. This should clarify the claim that a possible world is not just a possible state that a system can show up in, but also a context of evaluation for the truth of a proposition.

While it is often the case that the world of narration, or speech event, is also the context of evaluation, *modal operators* are grammatical categories and/or linguistic expressions that serve the function of shifting the context of evaluation from *this* world to other worlds. For example, if I say "John *would have* bought the bicycle," I am presupposing that in *this* world (the actual world that obtains at the time of speaking), it is false that John bought the bicycle; and I am asserting that there is some other world (say, one in which he had enough money for the purchase), in which it is true that he bought the bicycle. Similarly, while it may be false that magic exists in *this* world, it is true that magic exists in the world of Harry Potter. (At least that is the case as described in the books by J. K. Rowling as they exist in *this* world; in other worlds, there may be other versions of Harry Potter in which magic is no less impossible than it is in our world.)

For present purposes, two particularly important modal operators are *may* and *must*. Loosely, and in the tradition of scholars like Lewis (1986), Kratzer (2012), and von Fintel and Heim (2011), to say "Sally must have broken into the house," is to say something like, in *all* worlds compatible with the evidence in *this* world (as it is known to *us*), it is true that Sally broke into the house. That is, it is necessarily the case that she broke into the house. Similarly, to say "Sally may have broken into the house," is to say something like, there exists *some* world, compatible with the evidence in *this* world, in which it is true that Sally broke into the house. That is, it is possibly the case that she broke into the house.

In other words, modal auxiliary verbs like *may* and *must* quantify over possible worlds compatible with a set of context-specific assumptions. Such assumptions, which are usually context-specific (and thereby shifting) and socially shared (however widely or narrowly), constitute a kind of epistemic ground that can license inferences and/or justify interpretations. In particular, such a ground contains not only relatively general beliefs about causal regularities (e.g., if someone's fingerprints are on an object, that person must have touched the object), but also relatively specific beliefs about particular circumstances (e.g., Sally's fingerprints were on the window of the house in question). And such beliefs thereby count as evidential commitments that shape and direct inferences, and hence mediate future beliefs. Section 8.5 will take up necessity and possibility, as well as the function and transformation of interpretive grounds, in greater detail.

Having discussed some of the logical and linguistic details of possible worlds, we may now turn to their quantitative aspects.

8.3 Probabilities of Possible Worlds

As used here, a probability measure is a function $P(\cdot)$ from worlds to non-negative real numbers, such that

$$\sum_{w \in \Omega} P(w) = 1 . \tag{8.11}$$

In other words, each world in a universe is assigned a probability and the sum of all such probabilities should equal 1. This should made sense, given the fact that, by definition, individual worlds are non-overlapping, and a universe consists of all possible worlds.

To represent the probability of proposition p, understood as a set of worlds, the same notation will be used, along with the following stipulation:

$$P(p) = \sum_{w \in p} P(w) . \tag{8.12}$$

In other words, the probability of a proposition is the sum of the probabilities of the worlds contained in that proposition, and hence the sum of the probabilities of the worlds in which the proposition is true.

The probability of a world w conditioned on proposition q may be calculated as follows:

$$P(w/q) = \begin{cases} 0 & w \notin q , \\ \frac{P(w)}{P(q)} & w \in q . \end{cases} \tag{8.13}$$

In other words, the probability of a world conditioned on the truth of a proposition is zero if the world is not in the proposition, and simply the probability of the world itself if the world is in the proposition—suitably renormalized, by dividing by $P(q)$, to reflect the fact that we are only quantifying over worlds in which the proposition is true (rather than the entirety of the universe per se).

Using this relation, the probability of proposition p, conditioned on another proposition q, may be readily calculated:

$$P(p/q) = \sum_{w \in p} P(w/q) = \sum_{w \in p \wedge q} \frac{P(w)}{P(q)}. \tag{8.14}$$

In other words, we simply sum over the conditional probabilities, $P(w/q)$, for each world w in p; and this is equivalent to summing over $P(w)/P(q)$, as per equation (8.13), for each world w in the intersection of p and q.

All that said, we usually work the other way. That is, rather than use the probabilities of possible worlds to calculate the probabilities of propositions, the probabilities and conditional probabilities of propositions, such as priors and likelihoods, will be used to calculate the probabilities of possible worlds. To see how this happens, let us abstract away from interpretants and consequents in the original symptom and sickness scenario, and thereby restrict our attention to signs and objects. Suppose, in particular, that a suitably truncated universe, which shall be labeled Ω', consists of the following worlds:

$$\Omega' = \{w_{11}, w_{12}, w_{21}, w_{22}\}, \tag{8.15}$$

where w_{ij} is the world in which $S(w_{ij}) = S_i$ and $O(w_{ij}) = O_j$.

Suppose, further, that the priors and likelihoods of this scenario are given by their usual formulations (which, as described in chapter 2, involve three adjustable parameters):

$$P(O) = \begin{bmatrix} P(O_1) & P(O_2) \end{bmatrix} = \begin{bmatrix} p & 1-p \end{bmatrix}, \tag{8.16a}$$

$$P(S/O) = \begin{bmatrix} P(S_1/O_1) & P(S_1/O_2) \\ P(S_2/O_1) & P(S_2/O_2) \end{bmatrix} = \begin{bmatrix} 1-a & b \\ a & 1-b \end{bmatrix}. \tag{8.16b}$$

Using the standard definition of conditional probability

$$P(S_i/O_j) = \frac{P(S_i \wedge O_j)}{P(O_j)}, \tag{8.17}$$

which is just a more familiar form of equation (8.14), these certainties can be used to calculate the *joint probability distribution*:

$$P(S \wedge O) = \begin{bmatrix} P(S_1 \wedge O_1) & P(S_1 \wedge O_2) \\ P(S_2 \wedge O_1) & P(S_2 \wedge O_2) \end{bmatrix}$$

$$= \begin{bmatrix} P(S_1/O_1) \cdot P(O_1) & P(S_1/O_2) \cdot P(O_2) \\ P(S_2/O_1) \cdot P(O_1) & P(S_2/O_2) \cdot P(O_2) \end{bmatrix} \tag{8.18}$$

$$= \begin{bmatrix} (1-a) \cdot p & b \cdot (1-p) \\ a \cdot p & (1-b) \cdot (1-p) \end{bmatrix}.$$

Insofar as $S_i \wedge O_j = \{w_{ij}\}$, such that each such *basic proposition* (in the joint probability distribution) consists of a set that contains only a single world, we have thereby calculated the probability of each possible world in this universe.

As may be seen using equation (8.18), and as should be expected (insofar as the union of such basic propositions constitutes the universe, Ω', in question), the sum of the probabilities of the worlds in the universe is equal to 1:

$$\sum_{ij} P(S_i \wedge O_j) = P(S_1 \wedge O_1) + P(S_1 \wedge O_2) + P(S_2 \wedge O_1) + P(S_2 \wedge O_2)$$

$$= \sum_{ij} P(\{w_{ij}\})$$

$$= \sum_{w \in \Omega'} P(w) \tag{8.19}$$

$$= P(w_{11}) + P(w_{12}) + P(w_{21}) + P(w_{22})$$

$$= (1-a) \cdot p + b \cdot (1-p) + a \cdot p + (1-b) \cdot (1-p)$$

$$= 1 .$$

Given the joint probability distribution, it is easy to compute distributions over smaller sets of variables through a process known as *marginalization*, whereby a dimension in such a distribution is summed out. Suppose, for example, that we want to know $P(S_1)$, the probability that the person does not evince the symptom. This can be found by the following calculation:

$$P(S_1) = \sum_{O \in dom(O)} P(S_1 \wedge O)$$

$$= P(S_1 \wedge O_1) + P(S_1 \wedge O_2) \tag{8.20}$$

$$= P(w_{11}) + P(w_{12})$$

$$= (1-a) \cdot p + b \cdot (1-p) .$$

We are now in a position to calculate the posterior probabilities, $P(O_i/S_j)$. In particular, we want to know the probability that proposition O_i is true, given the fact that proposition S_j is true. To do this, we first find that subset of the universe that consists of all worlds compatible with S_j, which in this case is simply $\{w_{j1}, w_{j2}\}$. We then ask, in such a sign-restricted subset of the universe: What is the probability that $O_i = \{w_{1i}, w_{2i}\}$ obtains? Recall equation (8.14). This is given by

$$P(O_i/S_j) = \frac{P(w_{ji})}{P(w_{j1}) + P(w_{j2})} = \frac{P(S_j \wedge O_i)}{P(S_j \wedge O_1) + P(S_j \wedge O_2)} . \tag{8.21}$$

Notice, then, how evidence works in this paradigm: *signs put restrictions on accessible regions of the universe, and hence on the probabilities of possible worlds*.

Having seen how to calculate the probabilities of possible worlds, and how to translate between possible worlds and propositions, let us return to the larger universe, the one that contains not only objects and signs, but also consequents and interpretants, in order to take into account both value and causality.

8.4 Bayesian Networks and Expected Value

Recall that Ω is the set of all worlds w_{ijkl} in the original symptoms and sickeness scenario. To calculate the probability of each world in this set, we need to calculate the joint probability distribution, using the *chain rule* for probabilities:

$$\begin{aligned} P(w_{ijkl}) &= P(C_i \wedge I_j \wedge S_k \wedge O_l) \\ &= P(C_i/I_j \wedge S_k \wedge O_l) \cdot P(I_j \wedge S_k \wedge O_l) \\ &= P(C_i/I_j \wedge O_l) \cdot P(I_j \wedge S_k \wedge O_l) \\ &= P(C_i/I_j \wedge O_l) \cdot P(I_j/S_k \wedge O_l) \cdot P(S_k \wedge O_l) \\ &= P(C_i/I_j \wedge O_l) \cdot P(I_j/S_k) \cdot P(S_k \wedge O_l) \\ &= P(C_i/I_j \wedge O_l) \cdot P(I_j/S_k) \cdot P(S_k/O_l) \cdot P(O_l) . \end{aligned} \tag{8.22}$$

As may be seen, in undertaking this derivation, we made the following two assumptions about conditional independence:

$$P(I_j/S_k \wedge O_l) = P(I_j/S_k) , \tag{8.23a}$$

$$P(C_i/I_j \wedge S_k \wedge O_l) = P(C_i/I_j \wedge O_l) . \tag{8.23b}$$

Crucially, equation (8.23a) does not mean that O does not influence I. It means that once an agent knows S, learning about O won't teach them anything that they did not already know about *I*. That is, *I* is independent of *O*, given *S*. Similarly, equation (8.23b) does not mean that *S* does not influence *C*. It means that once an agent knows *I* and *O*, learning about *S* will not teach them anything that they did not already know about *C*. That is, *C* is independent of *S*, given *I* and *O*.

Framed in another way, and focusing on the last line in equation (8.22), we have been implicitly assuming that the causal network underlying the scenario is configured in such a way that the following information is all that is needed to calculate the joint probability distribution: $P(O)$, $P(S/O)$, $P(I/S)$, and $P(C/I \wedge O)$. Such assumptions are very general, in that they underlie not just the analysis of this simple scenario (allowing us to calculate its joint probability distribution), but also the main equation (from chapter 1). While they are easily modified or abandoned, it is arguably the case that many semiotic processes of the canonical sort take such assumptions for granted.

If we were to draw the *causal graph* (or Bayesian network) that embodies these simplifying assumptions, it has the following structure:

$$O \rightarrow S \rightarrow I \rightarrow C, \tag{8.24a}$$

$$O \rightarrow C. \tag{8.24b}$$

In particular, such a network contains three *random* variables (O, S, C) and one *decision* variable (I). See figure 8.1, which shows the causal relations among these four variables (with large vectors) in relation to the semiotic agent, and the *slash* that separates the variables that can be directly sensed and instigated by this agent from the variables that the agent can only indirectly project (or infer) and effect (or create).

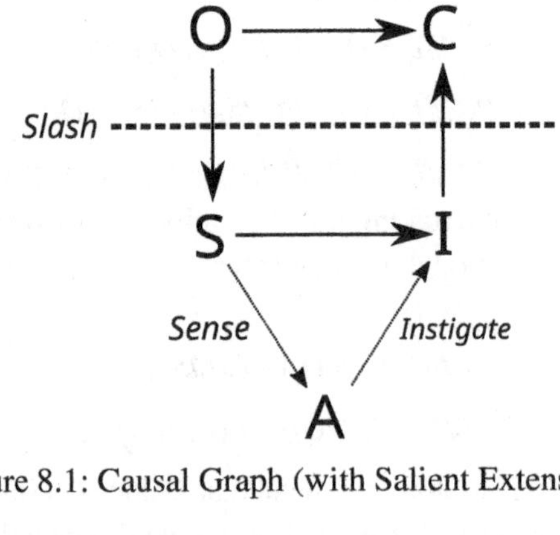

Figure 8.1: Causal Graph (with Salient Extensions)

Crucially, in distinguishing between decision variables and random variables, we are distinguishing between those variables that the agent can directly alter (through its instigations), and those variables that it can—at best—only alter indirectly (insofar as their probabilities are typically determined ahead of time, if only conditionally so). To be sure, the probability that an agent instigates a certain interpretant in the context of a particular sign is itself determined by such certainties, in conjunction with the agent's values. Recall, for example, the method of calculating $P(I/S)$ given $V(I/S)$, as introduced in chapter 6.

Before concluding this section, it is worthwhile using the foregoing analysis to express the main equation as a summation over possible worlds (subject to particular constraints):

$$\begin{aligned} V(I_i/S_j) &= \sum_{kl} V(C_k \wedge I_i \wedge O_l) \cdot P(C_k/I_i \wedge O_l) \cdot P(O_l/S_j) \\ &= \sum_{kl} V(C_k \wedge I_i \wedge O_l) \cdot P(C_k/I_i \wedge O_l) \cdot P(I_i/S_j) \cdot P(O_l/S_j) \cdot P(S_j) \\ &= \sum_{kl} V(C_k \wedge I_i \wedge O_l) \cdot P(C_k/I_i \wedge O_l) \cdot P(I_i/S_j) \cdot P(S_j/O_l) \cdot P(O_l) \\ &= \sum_{kl} V(C_k \wedge I_i \wedge S_j \wedge O_l) \cdot P(C_k \wedge I_i \wedge S_j \wedge O_l) \\ &= \sum_{kl} V(w_{kijl}) \cdot P(w_{kijl}) \\ &= \sum_{w \in I_i \wedge S_j} V(w) \cdot P(w). \end{aligned} \tag{8.25}$$

In this derivation, we have used Bayes' equation; the derivation of equation (8.22) in reverse; the fact that $P(S_j) = 1$, insofar as that is the sign on which the agent is conditioning; and the fact that $P(I_i/S_j) = 1$, insofar as that is the interpretant-sign relation whose value is being calculated.

As may be seen, the main equation, as introduced in chapter 1 and used throughout this book, is simply the expected value of an interpretant, conditioned on a sign, as summed over all possible worlds, subject to certain constraints. In this sense, the main equation is even simpler than it initially appeared.

The next section analyzes the sets of propositions that agents take for granted, as well as the way that such propositions mediate various kinds of necessity and possibility for those agents.

8.5 Conversational Backgrounds

Kratzer (2012) introduced the influential notion of a *conversational background*, denoted as $f(\cdot)$, which may be understood as a mapping from worlds (in some universe) to sets of propositions (where each proposition is itself a set of worlds, and thus a subset of the universe in question):

$$f(w) = \{p_1, p_2, \ldots\}. \tag{8.26}$$

Conversational backgrounds are thus world-dependent premise sets, where a set of premises is similar to a set of beliefs that an agent holds or a set of propositions that an agent is committed to.

While Kratzer focused on world-dependent premise sets, the idea of a conversational background may be readily extended to context-dependent premise sets. Such a dependence on context implies that the premises in question are not just collectivity-specific and agent-specific, but also time-dependent. Using the terminology of Roman Jakobson and, before him, Otto Jespersen, conversational backgrounds are thus *shifters* insofar as they are sensitive to the event of speaking and/or the act of interpretation.

Building on her notion of a conversational background, Kratzer introduced relatively general definitions of necessity and possibility. In particular, proposition p is *simple f-necessary* in world w if and only if

$$\cap f(w) \subset p, \tag{8.27}$$

meaning that p is true in every world in the intersection of the premise set of w. In other words, p follows from $f(w)$. Loosely, given an agent's beliefs in some world (where such beliefs are to be understood as a set of propositions), a rational agent is compelled to believe p (insofar as this proposition logically follows from the propositions, qua beliefs, that the agent is already committed to). See figure 8.2, in which *all* worlds in the intersection of the propositions (denoted by dashed lines) in the premise set are contained in proposition p (denoted by a solid line).

Similarly, proposition p is *simple f-possible* in world w if and only if

$$\cap \left(f(w) \cup \{p\}\right) \neq \emptyset. \tag{8.28}$$

That is, there is at least one world in the intersection of the union of p and the premise set of w. In other words, p is compatible with $f(w)$. Loosely, given an agent's beliefs in some world (understood as a set of propositions), a rational agent is not compelled to believe $\neg p$ (insofar as this proposition is not logically ruled out by the propositions that the agent is already committed to). See figure 8.3, in which there is *at least one* world in the intersection of the union

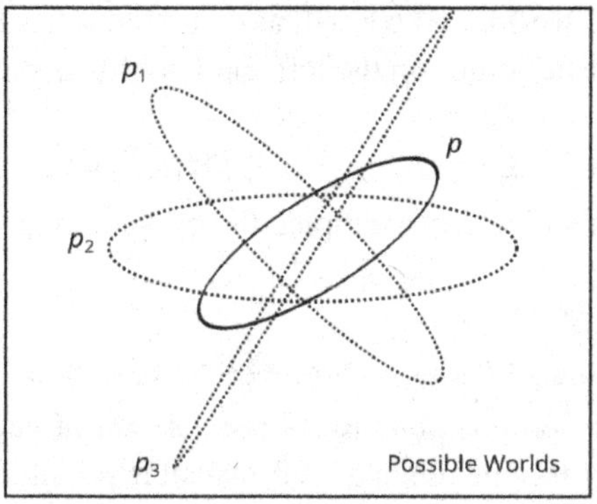

Figure 8.2: Proposition p is simple f-necessary in w, where $f(w) = \{p_1, p_2, p_3\}$

of proposition p (denoted by a solid line) and the propositions (denoted by dashed lines) in the premise set.

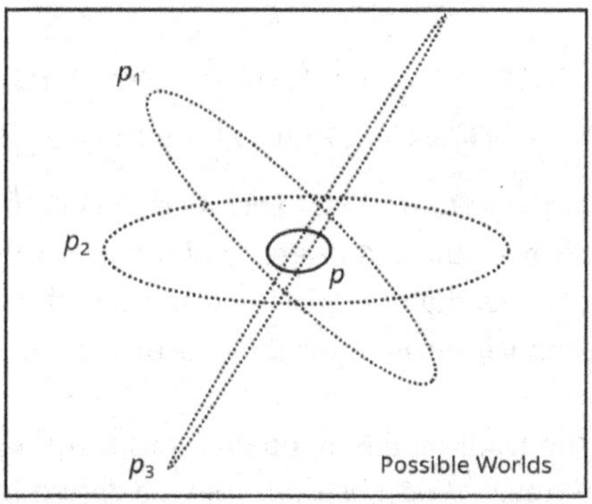

Figure 8.3: Proposition p is simple f-possible in w, where $f(w) = \{p_1, p_2, p_3\}$

As agents add more and more propositions to their premise sets (by collecting evidence, making inferences, undertaking actions, and thereby changing circumstances), more and more worlds get ruled out, such that the set of still-plausible possible worlds gets smaller and smaller for an agent. (To be sure, as will be taken up at length in chapter 9, agents may also adjust and abandon the propositions that they already have, as well as add others, in the face of contradictory evidence and in response to creative hypotheses.) To take into account this temporal dependence, we will often make f an explicit function of time as well as world, and we will assume that it has the following kind of structure (with earlier times on the left and later times on the right):

$$\cap f(t_0, w) \supset \cap f(t_1, w) \supset \ldots \supset \cap f(t_{n-1}, w) \supset \cap f(t_n, w) . \qquad (8.29)$$

In the initial scenario, for example, there were five key times, with the following order:

$$t_{\text{setup}} < t_{\text{sensation}} < t_{\text{inference}} < t_{\text{instigation}} < t_{\text{creation}} . \qquad (8.30)$$

By adding further propositions to its premise set at each point in time, the agent thereby narrows the range of still-plausible possible worlds: establishing a scenario, sensing a sign, making an inference, instigating an interpretant, and effecting a consequent.

To see how the foregoing issues play out in a simple scenario (that does not involve probabilities), let us return to the truncated universe $\Omega' = \{w_{11}, w_{12}, w_{21}, w_{22}\}$, where $w_{ij} = S_i \wedge O_j$. Suppose that we have the following two propositions:

$$p = S_2 = \{w \mid S(w) = S_2\} = \{w_{21}, w_{22}\}, \qquad (8.31a)$$

$$q = O_2 = \{w \mid O(w) = O_2\} = \{w_{12}, w_{22}\}. \qquad (8.31b)$$

Given the way that the ontology was specified, p is equivalent to the proposition "The person evinces the symptom," and q is equivalent to the proposition "The person is sick." As may be seen, and in line with the preceding discussion, each such proposition is equivalent to the set of worlds in which it is true.

Depending on the truth or falsity of such propositions, the truth value of a range of more complicated propositions may be determined as follows:

w	p	q	$p \wedge q$	$\neg(p \vee q)$	$p \Rightarrow q$	$(p \Rightarrow q) \wedge p$
w_{11}	0	0	0	1	1	0
w_{12}	0	1	0	0	1	0
w_{21}	1	0	0	0	0	0
w_{22}	1	1	1	0	1	1

Alternatively, collecting all the worlds (on the left-hand side) in which each of the propositions (across the top) is true, we have the following sets of worlds corresponding to each of these propositions:

$$p=\{w_{21},w_{22}\}, \tag{8.32a}$$

$$q=\{w_{12},w_{22}\}, \tag{8.32b}$$

$$p\wedge q=\{w_{22}\}, \tag{8.32c}$$

$$\neg(p\vee q)=\{w_{11}\}, \tag{8.32d}$$

$$p\Rightarrow q=\{w_{11},w_{12},w_{22}\}, \tag{8.32e}$$

$$(p\Rightarrow q)\wedge p=\{w_{22}\}. \tag{8.32f}$$

To return to the notion of an interpretive ground as it was introduced in chapter 1, $p\Rightarrow q$ is equivalent to the proposition "If the person evinces the symptom, then the person is sick." In effect, it is an assumption about causal relations, or at least correlations, and hence it is an assumption that may thereby license semiotic processes. Note, then, that a conversational background may be understood as a logical and linguistic framing of an interpretive ground, where the latter may be understood as the sensibilities and assumptions that agents have regarding the relations between different kinds of entities and events (and, at least for Charles S. Peirce, the ways that these relations may turn on shared conventions, causal processes, similar qualities, and the like). While a much narrower notion, to be sure, a conversational background has the advantage of being relatively easy to render mathematically.

Having reviewed some of the relations between propositions, worlds, truth tables, and various kinds of grounds, let us return to Kratzer's way of framing necessity and possibility. Suppose that we have a simple conversational background f, whose premise set at time t_1 contains only the following proposition (regarding causal relations):

$$f(t_1,w)=\{p\Rightarrow q\}=\{\{w_{11},w_{12},w_{22}\}\}. \tag{8.33}$$

As may be seen, if we were to take the union of $\{p\}=\{\{w_{21},w_{22}\}\}$ and this premise set, the intersection of that set of propositions would be $\{w_{22}\}$. In other words, $\cap(f(t_1,w)\cup\{p\})\neq\emptyset$. Following the definition given here, p is thereby f-possible in w (at t_1). A similar argument would show that $\neg p$ is also f-possible in w (at t_1), as are both q and $\neg q$. In other words, none of these propositions have yet been ruled out.

Suppose that at some later time t_2, proposition p is added to the conversational background, such that f is now given by:

$$f(t_2,w)=\{p\Rightarrow q,p\}=\{\{w_{11},w_{12},w_{22}\},\{w_{21},w_{22}\}\}. \tag{8.34}$$

As may be seen, the intersection of this premise set is $\{w_{22}\}$. And it is definitely the case that q is true in all worlds that are in this intersection: $\cap f(w) \subset q$. Following the definition given here, q is thereby f-necessary in w (at t_2). A similar argument would show that $\neg q$ is not f-possible in w (at t_2). Phrased in another way, having learned that the person evinces the symptom (p), the agent concludes that the person is definitely sick (q).

This change in f, from before the introduction of p to after the introduction of p, and hence in the movement from t_1 to t_2, may be understood as the updating of a premise set due to evidence, or the updating of an interpretive ground due to the sensing of a sign. In particular, all the agent assumes at first is the causal relation, so both q and $\neg q$ are possible. After it senses a sign, it adds to its set of assumptions the fact of the sign itself (p), so q becomes necessary (and $\neg q$ becomes impossible). In this way, the antecedent of the conditional is the sign-event (functioning like a minor premise), the consequent of the conditional is the object-event (functioning like a conclusion), and the conditional itself is the ground (functioning like a major premise).

To return to the discussion at the end of section 8.2, it should be clear that, in many possible worlds, semiotic processes are inherently modal in an epistemic sense: they inform agents what may be the case and what must be the case, given their evidence (qua sign), and given some causal (or at least implicational) understanding of the environment (qua ground). As should also be clear, in many possible worlds semiotic processes are also inherently modal in a deontic sense: they tell agents what they should and should not do, given the normative, legal, economic, epistemic, and/or ethical demands that they are subject to.

Before fleshing out the temporal and modal dynamics of a semiotic scenario in full, let us turn to similar dynamics in a universe that admits mathematical functions of a more stereotypical sort.

8.6 Worlds, Times, and Scales

Suppose that we have an ontology with four variables (M, B, X, Y), each of which has the set of real numbers as its domain of values:

$$\mathrm{dom}(M) = \mathrm{dom}(B) = \mathrm{dom}(X) = \mathrm{dom}(Y) = \mathbb{R}\,. \tag{8.35}$$

Let w_{mbyx} be the possible world in which M takes the value m, B takes the value b, Y takes the value y, and X takes the value x. Also, let W' be the set of possible worlds that are specifiable in this way:

$$W' = \{w_{mbxy} \mid m, b, x, y \in \mathbb{R}\}. \tag{8.36}$$

In effect, every point in $\mathbb{R}^4$, understood as a four-dimensional space that goes on forever in every direction, is a possible world in this universe. As intimated by the names of these variables, a simple structure is being imposed on this universe, one involving linear functions (along with their slopes M and y-intercepts B) that map values of variable X onto values of variable Y:

$$y = m \cdot x + b\,. \tag{8.37}$$

Suppose that a time-dependent conversational background f with the following characteristics is now added. First, as per the discussion in section 8.5, worlds compatible with the premise set at any point in time are a subset of the worlds that were compatible with the premise set at earlier points in time. Second, there are four key moments in time when the background transforms: $t_0 < t_1 < t_2 < t_3$. In particular, at time t_0, we have what Kratzer would call an *empty* conversational background (in which anything goes, as it were). At time t_1, a linear function is imposed on possible worlds. At time t_2, the values of the parameters in this function are set. Finally, at time t_3, the value of the input to this parameterized function becomes known.

Suppose, in particular, that the conversational background changes over time in the following way:

$$f(t_0, w) = \emptyset\,, \tag{8.38a}$$

$$f(t_1, w) = \{y = m \cdot x + b\}, \tag{8.38b}$$

$$f(t_2, w) = \{y = m \cdot x + b,\ m = 3,\ b = 1\}, \tag{8.38c}$$

$$f(t_3, w) = \{y = m \cdot x + b, m = 3,\ b = 1,\ x = 2\}. \tag{8.38d}$$

From these transformations, it should be clear that at time t_0, any world is possible, so long as it is contained in the original universe W'. In effect, the world that obtains might be located anywhere in $\mathbb{R}^4$. At time t_1, the universe becomes three-dimensional: the value of any three variables may be chosen at will, but then the value of the fourth variable is necessarily determined. At time t_2, the proposition $y = 3 \cdot x + 1$ becomes simple f-necessary. All possible worlds lie on this line. Finally, at time t_3, there is only one possible world in the universe: w_{3127}. All other worlds in the original universe have become, as it were, simple f-impossible.

Alternatively, we might have originally considered universe W, where $W \subset W'$, consisting of the following worlds:

$$W = \{w_{xy} \,|\, y = m \cdot x + b,\ m = 3,\ b = 1;\ x, y \in \mathbb{R}\}. \tag{8.39}$$

In effect, we have taken the linearization and parameterization for granted from the very beginning, and thereby limit our attention to a subset of the larger universe—one with much simpler temporal dynamics.

Indeed, we might even have considered universe W'', where $W \subset W' \subset W''$, consisting of the following worlds:

$$W'' = \{w_{nmbxy} \mid y = n \cdot x^2 + m \cdot x + b;\ n, m, b, x, y \in \mathbb{R}\}. \tag{8.40}$$

In effect, we turn our attention to an even larger universe—one in which the other two universes were simply smaller regions. Such upward ($W' \rightarrow W''$) and downward ($W' \rightarrow W$) movements show how easy it is to *reframe* and/or *rescale* universes as a function of the interests of the analyst or the inferences of the agent.

Having shown how the foregoing considerations can apply to mathematical functions and temporal unfoldings, and having shown how analysts and agents may easily move across more-or-less encompassing universes as sets of worlds, let us return to the original scenario.

8.7 Cutting the Universe Down to Size

As should be clear, the semiotic agents in this mathematical model of meaning, at least the rational ones analyzed in Part I, could also be framed as having a set of context-specific assumptions that they take for granted, a set of assumptions that we have been calling their *interpretive ground*. While not a conversational background in the strict sense, such an interpretive ground can certainly be understood as a set of relatively backgrounded premises, or presuppositions, that semiotic agents depend on—premises that enable and constrain their semiotic and social practices as much as their cognitive and computational processes. In the case of the symptoms and sickness scenario introduced in chapter 2, this ground was constituted by a set of commitments $\{p_1, p_2, ..., p_n\}$ that turned on the existence of the following:

- An ontology, consisting of four variables and their possible values:

$$C = \{C_1, C_2\},\ I = \{I_1, I_2\},\ S = \{S_1, S_2\},\ O = \{O_1, O_2\}$$

- A set of parameters, along with the values that they were set to:

$$p = .5,\ a = .1,\ b = .1,\ r = .7,\ q = .3,\ c = .2$$

- Various parameter-sensitive relations among such variables:

$$P(O),\ P(S/O),\ P(C/I \wedge O),\ V(C \wedge I \wedge O)$$

- The main equation, Bayes' equation, and some principle of evaluation:

  equation (1.1), equation (1.2), equation (2.33).

In every world in Ω, at every moment of time, the agent took this set of assumptions for granted. Such a ground literally constituted the *worldsview* of the agent: its view of the possible worlds (plural) in its universe, along with their characteristics, probabilities, values, and causal interrelations, in relation to its own sensations, instigations, and interests. In effect, and playing somewhat fast and loose with the formalism, we have

$$f(w) = \{p_1, p_2, \ldots, p_n\} = \textit{the ground}. \tag{8.41}$$

Crucially, as the agent gathered evidence (by sensing signs) and undertook actions (by instigating interpretants), it added extra assumptions to its ground—all of which constrained how the universe could be. For example, as soon as it sensed a sign, 8 of the 16 original possible worlds in its universe were determined to be impossible. Similarly, as soon as it instigated an interpretant, 4 of the 8 remaining worlds were no longer possible. See Appendix B for a step-by-step demonstration of this. By sensing and instigating in this way, the agent literally cut its universe down to size.

More carefully, insofar as the agent's premise set changed over time as a function of what it sensed, inferred, and instigated, such a background required a temporal index in addition to a world index. In particular, there were at least four times that had to be taken into account: (1) the time before the sign was sensed, when all the agent had was the original premise set, qua ground; (2) the time after the sign was sensed and before the inference was made (when the agent added S_j to its premise set); (3) the time after the inference was made and before the interpretant was instigated (when the agent added $P(O/S_j)$ to its premise set); (4) the time after the interpretant was instigated (when the agent added I_i to its premise set); and so forth. Phrased in another way, and echoing what was said in earlier sections of this chapter, the agent's *changing* ground was really a function from worlds and times to premise sets as follows:

$$f(t_0, w) = \{p_1, p_2, \ldots, p_n\}, \tag{8.42a}$$

$$f(t_{\text{sign}}, w) = \{p_1, p_2, \ldots, p_n, S_j\}, \tag{8.42b}$$

$$f(t_{\text{inference}}, w) = \{p_1, p_2, \ldots, p_n, S_j, P(O/S_j)\}, \tag{8.42c}$$

$$f(t_{\text{interpretant}}, w) = \{p_1, p_2, \ldots, p_n, S_j, P(O/S_j), I_i\}. \tag{8.42d}$$

As may be seen, the most salient part of what is f-necessary at time t_{sign} is added to the premise set at time $t_{\text{inference}}$, and the most salient part of what is f-necessary at time $t_{\text{inference}}$ is added to the premise set at time $t_{\text{interpretant}}$. In short, not only did the agent use its equations, parameters, and inputs to determine which propositions were no longer simple f-possible, and thus which worlds were now impossible, it also determined which propositions were simple f-necessary, and hence which propositions it must believe and which actions it must undertake. Indeed, continuing in this vein, part of what is f-necessary at time $t_{\text{interpretant}}$ is the probabilities of the consequents, given the interpretant and the (updated) probabilities of the objects: $P(C/I_i \wedge O)$. The process could thereby continue.

We have so far been focused on the ways in which an agent can narrow down a universe into smaller and smaller sets of possible worlds. To be sure, as shown in section 8.6, analysts and agents often work the other way. In the current scenario, for example, the agent assumes that there are 16 possible worlds, whose initial probabilities, desirabilities, and interrelations are already determined. We could imagine a more encompassing scenario in which the set of worlds includes all possible parameterizations of the same scenario. Indeed, as analysts, we actually explored the agent's behavior in differently parameterized versions of the same scenario. And, in the next chapter, we will even analyze scenarios in which a meta-agent has to find out what world it is in within this larger universe: not only which of the 16 possible worlds, but also which of the infinitely possible parameterizations (of the probabilities, and perhaps even desirabilities, of those worlds). In these ways, we often tack between the following two intimately related universes:

$$\Omega = \{w_{ijkl} \mid i,j,k,l \in \{1,2\}\}, \tag{8.43a}$$

$$\Omega' = \{w_{pabrqcijkl} \mid p,a,b,r,q,c \in [0, 1];\ i,j,k,l \in \{1,2\}\}. \tag{8.43b}$$

As analysts, we can bump up into even larger universes: the universe of all possible scenarios that are specifiable with the original model (a few of which were explored in earlier chapters); or the universe of all possible models of possibly specifiable scenarios (where agents act under different sets of constraints altogether).

Several more regions of these larger universes will be taken up in the next chapter.

9 Meta-Semiotic Processes

Overview

As simple as the main equation from chapter 1 was, it took into account multiple modes of mediation: not just interpretant-sign mediation (through the agent and its values and commitments), but also object-sign and interpretant-consequent mediation. This chapter uses that equation, in a variety of extended forms, to model meta-semiotic processes. Section 9.1 shows the relation between the main equation and a variety of closely related equations that may be used to model scenarios with greater and fewer modes of mediation, including those that arise due to the presence of additional agents. Section 9.2 surveys the variety of ways that agents may transform their models of the world over time, and it thereby shows the ways that interpretive grounds may be updated by relatively self-reflective agents. Section 9.3 shows how to adjust the main equation to model self-updating agents that improve the fit between their ground and the world by way of meta-semiotic processes. Section 9.4 shows how to adjust the main equation to model meta-semiotic agents that take into account the interpretive grounds of other agents. Section 9.5 discusses the relation between the foregoing issues and culture, as a kind of intersubjectively held interpretive ground. And the conclusion examines a variety of mediating agents—such as enemies, parasites, translators, and noise—that may stand between a signer and an interpreter. As will be seen, to fully flesh out all these considerations would require another book. We will therefore content ourselves with surveying this vast terrain and highlighting some of the mathematical complexities that arise within it.

9.1 Variations on the Main Equation

In the introduction to this book, we offered both a key diagram and the main equation, where each could be understood as an interpretant (or object) of the other as a sign. In this section, we offer a series of diagram-equation relations that both lead to, and follow from, that first diagram-equation relation. This is done, in part, to better explicate the logic of the original diagram-equation relation and to show the relation between this mathematical model of meaning and a variety of other models that are either incorporated by it or incorporating of it.

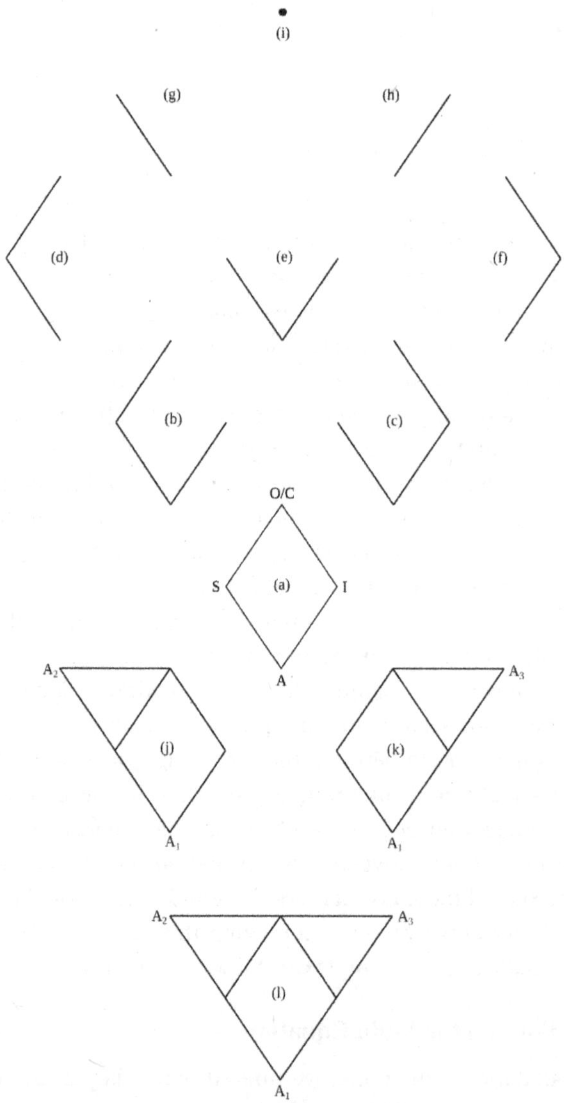

Figure 9.1: Incorporated and Incorporating Diagrams

Focus first on part (a) of figure 9.1. As introduced in chapter 1, and explored in subsequent chapters, this diagram represents an agent (A) that not only senses signs (S) and instigates interpretants (I), but also projects objects (O) beyond what it senses and effects consequents (C) beyond what it instigates. Governing these relations was the main equation, which turned on the agent's priors, likelihoods, eventualities, and values. It is reproduced here for ease of comparison:

$$V(I_i/S_j) = \sum_{kl} V(C_k \wedge I_i \wedge O_l) \cdot P(C_k/I_i \wedge O_l) \cdot P(O_l/S_j)\,. \tag{9.1}$$

In effect, this diagram-equation relation is a model of the *sensing, projecting, instigating, and effecting* agent. It has, in a wide variety of guises, been the central focus of our analysis and efforts so far.

A range of other modes of mediation are now easy to imagine. As seen in part (b) of figure 9.1, and as explored in sections 4.3, 4.4, and 6.3, the agent may not take into account interpretant-consequent mediation; or consequents may be so rigidly determined by interpretants that the mapping between them is essentially the identity matrix. What the agent instigates *is* what there is (or will be), as it were. This diagram would be paired with the following equation:

$$V(I_i/S_j) = \sum_{l} V(I_i \wedge O_l) \cdot P(O_l/S_j)\,. \tag{9.2}$$

In effect, this diagram-equation relation is a model of the *sensing, projecting, and instigating* agent. If value is replaced with utility, it may be usefully compared with Savage's (1972) classic account of decision-making under uncertainty.

Conversely, as seen in part (c) of figure 9.1, the agent may not take into account object-sign mediation, or signs may be so rigidly determined by objects that the mapping between them is essentially the identity matrix. What the agent senses *is* what there is (or was), as it were. This diagram would be paired with the following equation:

$$V(I_i/S_j) = \sum_{k} V(C_k \wedge I_i) \cdot P(C_k/I_i \wedge S_j)\,. \tag{9.3}$$

In effect, this diagram-equation relation is a model of the *sensing, instigating, and effecting* agent. If value is replaced with utility, it resembles classic models of mediated action (in which the sign plays the role of context, the interpretant plays the role of the means, and the consequent plays the role of the end).

Having removed one mode of mediation from the main diagram and equation, we may now remove two modes of mediation. As seen in part (d) of figure 9.1, the agent may overlook both interpretant-consequent mediation and sign-interpretant mediation. In effect, the agent infers, projects, or otherwise comes to know an object as mediated by a sign, and that is that. While its priors may turn into posteriors as it updates its knowledge, the model does not take into account how the agent might eventually act on that knowledge. As such, there is no mediation by values in the strict sense, as there are no actions to be undertaken (yet) that turn on such values. This diagram would be paired with the following equation, which is simply Bayes' equation (which itself is already implicitly present in the last term of the main equation):

$$P(O_l/S_j) = \frac{P(S_j/O_l) \cdot P(O_l)}{\sum_m P(S_j/O_m) \cdot P(O_m)} . \tag{9.4}$$

This diagram-equation relation is a simple model of the *sensing and projecting* agent. It is equivalent to classic accounts of Bayesian inference so long as signs are treated as 'evidence' and objects are treated as 'hypotheses'.

As seen in part (e) of figure 9.1, the agent may not take into account both object-sign mediation and interpretant-consequent mediation. In effect, the agent resides in a world of pure immediacy: what it senses *is* what is (or was); what it instigates *is* what is (or will be). Alternatively, for readers wary of immediacy, it can mean that signs correlate perfectly with objects and consequents correlate perfectly with interpretants. In effect, the agent simply instigates the most valuable interpretant, given the sign that they sense. This diagram would be paired with the following equation:

$$V(I_i/S_j) = V(I_i \wedge S_j) . \tag{9.5}$$

This diagram-equation relation is a simple model of the *sensing and instigating* agent. The agent does (I) what it deems best to do, in light of its values and given the circumstances (S).

As seen in part (f) of figure 9.1, the agent may not take into account both object-sign mediation and sign-interpretant mediation. In effect, the agent instigates the interpretant that will produce the most valuable consequent, given its values and commitments. This diagram would be paired with the following equation:

$$V(I_i) = \sum_k V(C_k \wedge I_i) \cdot P(C_k/I_i) . \tag{9.6}$$

This diagram-equation relation is a simple model of the *instigating and effecting* agent. If I is framed as the means and C as the end, it offers a simple example of circumstance-independent (or context-free) instrumental action.

Three more diagrams may now be added that are at the edges of this approach. Part (g) of figure 9.1 shows an *affected* agent. It can sense, or be affected, and thereby register changes in its environment, but that is all. Part (h) of figure 9.1 shows an *affecting* agent. It can instigate or have effects that are registered in the environment as changes, but that is all. Finally, there is part (i), understood as a lone point as opposed to a mediating line, representing that which is neither affected nor affecting. While not a *monad* in a strict sense, it is certainly isolated and unreactive. Note, crucially, that such affected and affecting agents are not to be contrasted with our original sensing, projecting, instigating, and effecting agent; rather, they should be understood as simpler modalities that are subsumed by it and/or contained within it. In effect, *affect* (in addition to effect) is at the heart of the model.

Besides moving towards more and more simple diagram-equation relations, in which modes of mediation are removed, we may also move towards more and more complicated diagram-equation relations, in which modes of mediation are added. While the mathematical part of these diagrams will not be taken up until section 9.5, it is useful to note their existence here and motivate their inclusion.

As was seen in part (a) of figure 9.1, it has been assumed that the agent assumes that the object-sign relation is relatively simple, in the sense that it is not mediated by another agent similar to itself (as modeled, say, by the main equation). If such an agent is taken into account, another triangle—or rather partially overlapping rhombus—must be added to the figure (see part (j) of figure 9.1, and also recall figure 1.3). In other words, to get to the object through the sign, the original agent (A_1) must take into account another agent (A_2)—one that reasons and evaluates much like itself. Phrased in another way, the sign that A_1 senses is itself the interpretant that A_2 instigates (and the sign that A_2 senses is, in effect, the object). In a Peircean formulation, the sign-object relation is now framed as a third (or mediator) as opposed to a second (or intermediary).

We have also been assuming that the agent assumes that the interpretant-consequent relation is relatively simple, in the sense that it is not mediated by another agent similar to itself (again, as modeled by the main equation). If such an agent is taken into account, another triangle—or, rather, a partially overlapping rhombus—must be added to the figure (see part (k) of figure 9.1). In other words, to get to the consequent through the interpretant, the original agent (A_1) must take into account another agent (A_3)—one that reasons and evaluates much like itself. Phrased in another way, the interpretant that A_1 instigates is itself the sign that A_3 senses (and the interpretant that A_3 instigates is, in effect, the consequent). In a Peircean formulation, the interpretant-consequent relation is now framed as a third as opposed to a second.

Finally, as seen in part (l) of figure 9.1, the main agent may assume that not only are the signs that it senses mediated by an agent similar to itself, so are the interpretants that it instigates. Loosely, there is an agent speaking to it, as well as an agent that it is speaking to. (Both of these agents, if not all three of them, could of course be one and the same agent, at different moments in time, as part of an ongoing dialogue or interaction, including an internal conversation.) Here, then, the agent must take into account two more agents, each of which reasons and evaluates like itself, using something like the main equation or one of its many variants. Putting both of the foregoing triangles together, the sign that A_1 senses is itself the interpretant that A_2 instigates; and the interpretant that A_1 instigates is itself the sign that A_3 senses.

In all of these last three cases, an agent's interpretive ground must take into account, or somehow incorporate, the interpretive grounds of other agents. In other words, its ontology, commitments, and values must make reference to the ontology, commitments, and values of one or more agents (whose grounds, reciprocally, may take into account its interpretive grounds). Crucially, once this level of mediation has been added, the process can embed and enchain indefinitely, with each agent taking into account the agency of the other, which may itself involve the ways that that agent takes into account their own agency, to greater or lesser degrees of adequacy, and so on, indefinitely.

We will explore many of these modes of mediation, and discover still others, in the sections that follow.

9.2 Fluid Grounds

In the scenarios analyzed in chapters 2 and 3, the agent's interpretive ground was relatively predetermined and fixed. In particular, not only were the priors $P(O)$, likelihoods $P(S/O)$, eventualities $P(C/I \wedge O)$, and values $V(C \wedge I \wedge O)$ already given (prior to any semiotic process that the agent might undertake), they also did not undergo any subsequent transformations (as a function of the success or failure of the agent's semiotic processes). To be sure, in later chapters we analyzed the way that interpretive grounds—in the guise of allele frequencies, register weights, or parameter values—get updated over generations or after interactions. But such scenarios turned on selected agents as opposed to selecting agents, and thus agents that evolve and learn rather than agents that choose. We are now interested in examining the ways that a relatively rational agent may adjust its interpretive ground as a function of its interpretive processes, and the feedback that it receives through its encounters with the environment—an environment that may include other agents like itself.

More specifically, of interest are the meta-signs that indicate that the agent's interpretive ground is out of touch with the world and/or in need of updating; and of interest are the meta-interpretants (of such meta-signs) that result in transformations of the agent's interpretive ground. In keeping with the diagrammatic imaginary offered in section 9.1, we are thereby interested in *self-reflective semiotic agents*. Such agents can have their own interpretive grounds, and hence their own priors, likelihoods, eventualities, and values, as both the objects and consequents of their meta-semiotic processes. See figure 9.2, which could just as easily have been the final part of figure 9.1.

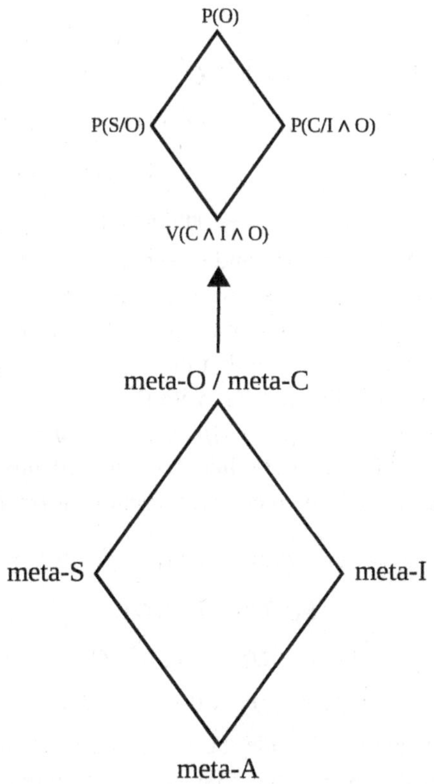

Figure 9.2: Self-Reflexive Semiotic Agent

This topic is, of course, enormous. See Kockelman (2017, 2020b) for a discussion of some of its intricacies in the context of Bayesian statistics and machine learning. In the rest of this section, we will simply lay out some of the key ways in which interpretive grounds may be transformed (building on the discussion at the end of chapter 8). Later sections will take up these issues in greater detail.

Interpretive grounds may be fluid, or transformable, in a wide range of ways. First, the contents of the agent's ontology may change. In particular, the sets of objects O, signs S, interpretants I, and/or consequents C may change in number and/or kind. For example, $O=\{O_1,O_2\}\rightarrow O'=\{O_1,O_2,O_3\}$, such that the set of objects changes in size. Similarly, $O_1=$ *person is upset* $\rightarrow O'_1=$ *person is livid*, such that some particular object in that set changes in character. In short, the cardinality and/or contents of such sets may change. These sorts of changes might be formalized as follows:

$$\begin{aligned} O=\{O_1,O_2,\ldots,O_n\} &\rightarrow O'=\{O_1,O_2,\ldots,O_{n'}\},\\ &\rightarrow O'=\{O'_1,O'_2,\ldots,O'_n\},\\ &\rightarrow O'=\{O'_1,O'_2,\ldots,O'_{n'}\},\\ &\rightarrow \text{and so on.} \end{aligned} \tag{9.7}$$

And it is similar for $S\rightarrow S'$, $I\rightarrow I'$, and $C\rightarrow C'$.

Second, the priors, likelihoods, eventualities, and values that interrelate and quantify such ontological contents may change. For example, $P(O)=[p,1-p]\rightarrow P'(O)=[p',1-p']$, such that the numerical values used in the priors are altered. Or, mixing this process with the first of the two processes just described, $P(O)=[p,1-p]\rightarrow P'(O')=[p',q',1-p'-q']$, such that not only are the prior probabilities of O_1 and O_2 altered, but a third alternative with its own probability is added as well. Such changes might be formalized as follows:

$$\begin{gathered} P(O)\rightarrow P'(O),\\ P(S/O)\rightarrow P'(S/O),\\ P(C/I\wedge O)\rightarrow P'(C/I\wedge O),\\ V(C\wedge I\wedge O)\rightarrow V'(C\wedge I\wedge O). \end{gathered} \tag{9.8}$$

Here, for simplicity, we leave aside the ways that parts of the ontology may change simultaneously, as per the first set of changes, such as $P(S/O)\rightarrow P'(S'/O')$, and so forth.

Third, the process whereby $V(I/S)$ is calculated, which depends on the foregoing items as its main arguments, may change. For example, the agent may switch from expected value to expected utility, or the agent may switch from

Bayesian reasoning and the main equation to some other form of artificial intelligence or means of calculation. Such changes might be formalized as follows:

$$\begin{aligned} V(I/S) = \text{Main Equation}\big(P(O), P(S/O), P(C/I \wedge O), V(C \wedge I \wedge O)\big) \\ \rightarrow \text{Another Equation (with Similar Arguments)}, \\ \rightarrow \text{Another Equation (with Other Arguments)}, \\ \rightarrow \text{Other Imaginaries, Rationalities, and Influences Altogether.} \end{aligned} \tag{9.9}$$

For the sake of simplicity, we again leave out all the ways that the first two sorts of changes might feed into, or otherwise influence, these changes.

Rather than explore all these possibilities—from mutable ontologies to transforming rationalities—the following section will focus on the second set of possibilities. In such meta-semiotic processes, as will now be shown, the meta-signs are co-occurrences of signs, objects, interpretants, and consequents; the meta-objects are the priors, likelihoods, and eventualities that constitute an agent's semiotic ground; and the meta-interpretants are the changes that such a meta-agent makes to the probabilities that constitute its priors, likelihoods, and eventualities.

9.3 Self-Updating Semiotic Agents

To model such meta-semiotic processes, whereby the grounds of lower-order semiotic processes become the objects of higher-order semiotic processes, we will continue to use the main equation, with each of its terms suitably adjusted. Things can get complicated very quickly at this level, so we will proceed slowly.

The first thing to adjust is the object. For present purposes, the meta-objects of these higher-order semiotic processes are the parameter values that underlie the priors, likelihoods, and eventualities of the agent's lower-order semiotic processes. Assuming a particularly simple ontology in which there are two objects, two signs, two interpretants, and two consequents, there are six such parameters. Recall, for example, the scenario examined in chapter 2 regarding symptoms and sickness. In particular, there is parameter p, which underlies the agent's priors:

$$P(O) = \begin{bmatrix} p & 1-p \end{bmatrix}. \tag{9.10}$$

There are parameters q and r, which underlie the agent's likelihoods:

$$P(S/O) = \begin{bmatrix} q & r \\ 1-q & 1-r \end{bmatrix}. \tag{9.11}$$

And there are parameters *t*, *s*, *u*, and *v*, which underlie the agent's eventualities:

$$P(C_1/I \wedge O) = \begin{bmatrix} s & t \\ 1-s & 1-t \end{bmatrix}, \tag{9.12a}$$

$$P(C_2/I \wedge O) = \begin{bmatrix} u & v \\ 1-u & 1-v \end{bmatrix}. \tag{9.12b}$$

In first-order semiotic processes, the values of all these parameters have relatively known, definite, and fixed values. In higher-order semiotic processes, in contrast, the values of these parameters are to be understood as uncertain, probabilistic, and alterable. In particular, rather than p being a discrete value that is known in advance and fixed (say, $p = .4$), we now have $P(p)$, understood as a continuous probability distribution, itself a function of p, which is updated over time. And it is similar for the other parameters: $q \rightarrow P(q)$, $r \rightarrow P(r)$, and so forth.

Imagine, for example, an agent that wants to make a bet that turns on the flip of a coin. Initially, the agent may assume that the coin is unbiased, such that $P(p) = 1$ for $p = 1/2$ (and zero otherwise). However, after watching a series of flips that seem to come up heads too often, the agent may assume that $P(p) = 2p$, meaning the coin is strongly biased towards heads (but precisely how much so is still undetermined).

The meta-signs in such higher-order semiotic processes will be conjunctions of the signs, objects, interpretants, and consequents of lower-order semiotic processes. For example, as a function of which kinds of objects occur (such as coins coming up heads or coins coming up tails), and how frequently, the agent will update $P(p)$, and hence the probability distribution organizing its priors. Likewise, as a function of which kinds of objects and signs co-occur, and how frequently, the agent will update $P(q)$ and $P(r)$, and hence the probability distributions organizing its likelihoods. And so on, and so forth.

Just as the grounds of lower-order semiotic processes involve priors and likelihoods, which are used by agents to determine posteriors through Bayes' equation, so do the meta-grounds of higher-order semiotic processes. In particular, the meta-priors are continuous probability distributions, for the parameters in question, prior to the agent's sensing of a meta-sign. The meta-likelihoods are the probabilities of the meta-signs given the meta-objects. And the meta-posteriors are the updated probability distributions, for the parameters in question, given whatever meta-sign has been sensed. The following table shows each of these constituents for all possible meta-objects in our simple scenario:

Sign	Object	Prior	Likelihood	Posterior
S	O	$P(O)$	$P(S/O)$	$P(O/S)$
O	p	$P(p)$	$P(O/p)$	$P(p/O)$
$O \wedge S_1$	q	$P(q)$	$P(O \wedge S_1/q)$	$P(q/O \wedge S_1)$
$O \wedge S_2$	r	$P(r)$	$P(O \wedge S_2/r)$	$P(r/O \wedge S_2)$
$C_1 \wedge I \wedge O_1$	s	$P(s)$	$P(C_1 \wedge I \wedge O_1/s)$	$P(s/C_1 \wedge I \wedge O_1)$
$C_1 \wedge I \wedge O_2$	t	$P(t)$	$P(C_1 \wedge I \wedge O_2/t)$	$P(t/C_1 \wedge I \wedge O_2)$
$C_2 \wedge I \wedge O_1$	u	$P(u)$	$P(C_2 \wedge I \wedge O_1/u)$	$P(u/C_2 \wedge I \wedge O_1)$
$C_2 \wedge I \wedge O_2$	v	$P(v)$	$P(C_2 \wedge I \wedge O_2/v)$	$P(v/C_2 \wedge I \wedge O_2)$

The first line of this table simply reviews the key components of first-order semiotic processes, as they have been used in preceding chapters. The seven lines that follow show how these components may be adjusted to capture higher-order semiotic processes. As may be seen by looking across the columns, the objects of higher-order processes are the parameter values of lower-order processes. The signs of higher-order processes are co-occurrences of the constituents of lower-order semiotic processes. And the priors, likelihoods, and posteriors of higher-order processes are conditional probability distributions that interrelate the meta-objects and meta-signs in question.

As always, Bayes' equation may be used to express posteriors in terms of likelihoods and priors. In the case of the lower-order semiotic processes analyzed in earlier chapters and summarized in the first line of the table, the following formulation of this equation was used:

$$\begin{gathered}\text{POSTERIOR} \propto \text{LIKELIHOOD} \times \text{PRIOR} \\ P(O_i/S_j) \propto P(S_j/O_i) \cdot P(O_i) \\ P(O_i/S_j) = \frac{P(S_j/O_i) \cdot P(O_i)}{\sum_k P(S_j/O_k) \cdot P(O_k)} .\end{gathered} \tag{9.13}$$

For the higher-order semiotic processes in the table, a different formulation of Bayes' equation will be used—one in which the normalizing summation over discrete values in the denominator becomes an integral over continuous values. For the meta-semiotic process summarized in the second line of the table, for example, we would have:

$$\text{META-POSTERIOR} \propto \text{META-LIKELIHOOD} \times \text{META-PRIOR}$$

$$P(p/O_i) \propto P(O_i/p) \cdot P(p)$$
$$P(p/O_i) = \frac{P(O_i/p) \cdot P(p)}{\int_0^1 P(O_i/p') \cdot P(p') \cdot dp'} . \tag{9.14}$$

Similar equations can be derived for each of the other meta-semiotic processes summarized in the table.

Before continuing, it is helpful to show how such posteriors might be evaluated and acted on using a higher-order version of the main equation. For the sake of simplicity, it will be assumed that there is no interpretant-consequent mediation in the higher-order semiotic processes, such that what the agent instigates is what ultimately occurs. As discussed in section 9.1, this means that the higher-order equivalent of equation (9.2) will be used rather than the higher-order equivalent of equation (9.1). It will also be assumed that the key interpretant of higher-order semiotic processes involves two steps. First, at the level of the meta-semiotic processes, the agent updates their old prior by replacing it with their new posterior (calculated through Bayes' equation, as just discussed). Second, the agent now uses the newly calculated prior (as a probability distribution) or the location of its maximum (as a discrete value) in their lower-order semiotic processes.

Continuing with the last example, the first step is relatively straightforward. The agent updates its probability distribution, such that $P(p) \rightarrow P'(p) = P(p/O_i)$, as given by equation (9.13). The second step is slightly more complicated, as its involves a higher-order version of the main equation:

$$\begin{aligned} V(p'/O_j) &= \int_0^1 V(p', p) \cdot P(p/O_j) \cdot dp \\ &= \int_0^1 \delta(p - p') \cdot P(p/O_j) \cdot dp \\ &= P(p'/O_j) . \end{aligned} \tag{9.15}$$

As may be seen in the first line of equation (9.15), several changes have been made. First, the usual summation over discrete values has been replaced by an integral over continuous values. Second, we now use the meta-posteriors calculated here: $P(p/O_j)$. Third, there is a continuous range of possible interpretants. In particular, just as the meta-sign is O_j, the meta-interpretant is now p', understood as the action of setting the value of p, which organizes the agent's first-order priors, to p' (insofar as that is understood to be the most valuable setting, given the meta-sign O_j).

As may be seen in the second line, the evaluation function $V(p',p)$ is being modeled as a Dirac delta function, $\delta(p-p')$, whose essential characteristics are as follows:

$$\delta(p-p')=\begin{cases}\infty & p=p' \\ 0 & p\neq p'\end{cases} \tag{9.16}$$

and

$$\int_{-\infty}^{\infty}\delta(p-p')\cdot dp=1\,, \tag{9.17}$$

such that

$$\int_{-\infty}^{\infty}\delta(p-p')\cdot f(p)\cdot dp=f(p')\,. \tag{9.18}$$

In short, and somewhat loosely, $\delta(p-p')$ is infinite when $p=p'$, equal to zero elsewhere, and suitably normalized such that its integral, over the real numbers, is unity. In effect, this function serves to translate a probability distribution $P(p/O_j)$, turning on a meta-object p, into an evaluative function $V(p'/O_j)$, turning on a meta-interpretant p'.

As usual, the agent should choose the most valuable interpretant p' given sign O_j. Given this choice of evaluative functions, it is assumed that no interpretant (qua value of p') is more costly than any other, so the agent should choose the most valuable value—which, as may be seen, is the same as the most probable value. In effect, the agent may simply use that p' at which $V(p'/O_j)=P(p'/O_j)$ is at its maximum.

Let us now move from agents that can update their own interpretive grounds to agents that can take into account the interpretive grounds of other agents.

9.4 Modeling Another Agent's Model

We now analyze a scenario in which there are two agents. Recall part (j) of figure 9.1, which is expanded—and slightly tweaked—as figure 9.3. Whereas one agent, A_1, senses signs and instigates interpretants, the other agent, A_2, senses objects and instigates signs. In particular, the signs instigated by A_2 are precisely the signs sensed by A_1, such that each agent is intimately coupled to the other, one doing the work of signification and the other doing the work of interpretation.

To work together in such a way, not only must each agent have a model of the world, but each agent's model of the world must now take into account the other agent's model. As will be seen, this scenario quickly leads to *recursive interpretive grounds*. For example, one agent can have a model of the other

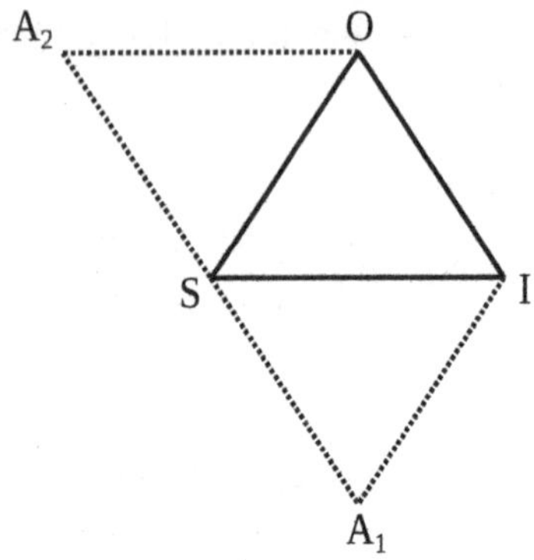

Figure 9.3: Modeling Another Agent's Model

agent's model of its own model. As will also be seen, this scenario allows us to model the relative mismatch (or fit) between models and meta-models.

Insofar as there are now two agents to take into account—the signifying agent (A_2) and the interpreting agent (A_1)—a subscript will be appended to each of the terms in the equations to indicate which agent's assumptions and values they are meant to model. Insofar as each agent takes into account the other agent's model (of its model (of their model (of its model))), and so on, a list of embedding indices will be used to indicate the relations in question. Thus, a term like $P_2(S/O)$ will denote A_2's model of the likelihood of S given O, whereas a term like $P_{1(2)}(S/O)$ will denote A_1's model of A_2's model of the likelihood of S given O. Similarly, a term like $V_1(I \wedge O)$ will denote A_1's model of the value of a world in which I and O co-occur, whereas a term like $V_{2(1)}(I \wedge O)$ will denote A_2's model of A_1's model of the value of a world in which I and O co-occur. And it is similar for other combinations, to any depth of recursive embedding.

In thinking about such meta-semiotic agents, the first simplifying assumption to make is that there is a one-to-one correspondence between interpretants and consequents, such that the latter can be elided from the analysis. Recall equation (9.2). The main equation, now suitably indexed to the agent in question, thereby becomes

$$\begin{aligned} V_1(I_i/S_j) &= \sum_k V_1(I_i \wedge O_k) \cdot P_1(O_k/S_j) \\ &= \sum_k V_1(I_i \wedge O_k) \cdot \frac{P_1(S_j/O_k) \cdot P_1(O_k)}{\sum_l P_1(S_j/O_l) \cdot P_1(O_l)} . \end{aligned} \tag{9.19}$$

As usual, Bayes' equation has been used to rewrite the posteriors $P_1(O/S)$ in terms of the likelihoods $P_1(S/O)$ and priors $P_1(O)$. As may be seen, these likelihoods encapsulate A_1's understanding of the patterning underlying A_2's behavior: the probability of sign S given object O.

Whereas A_1 senses signs and instigates interpretants, A_2 senses objects and instigates signs, all the while trying to predict what interpretant A_1 will offer of its sign. This agent's behavior will be modeled using a variant of the main equation:

$$V_2(S_i/O_j) = \sum_k V_2(I_k \wedge O_j) \cdot P_2(I_k/S_i)\,. \tag{9.20}$$

As may be seen, $P_2(I/S)$ encapsulates A_2's understanding of the patterning underlying A_1's behavior: the probability of interpretant I given sign S.

In short, just as $V_1(I/S)$ makes reference to $P_1(S/O)$, $V_2(S/O)$ makes reference to $P_2(I/S)$: each agent has assumptions about the other agent's behavior, even if it does not yet recognize the other as an evaluative and rational agent per se. With these two equations, we may examine the details of each agent's model at deeper levels of recursion.

But let us not go deeper just yet. Each agent could, of course, simply treat the other agent as a non-agent, and thereby model its behavior as a relatively straightforward likelihood, be it $P_2(I/S)$ or $P_1(S/O)$. In effect, each agent would treat the other as if its behavior were generated by causal propensities (in the world), or embodied tendencies (such as an unconscious habit), rather than by a mindful agent much like itself, whose behavior would thus also be guided by an inferential and evaluative process. These two equations *as such* could now be used to model the joint behavior of the two agents: what actually ends up happening, given their assumptions about the world, and the way that their behaviors are inherently coupled. In effect, each agent would act agentively, while treating the other agent as if it acted non-agentively—or at least less than fully agentively.

This is a fair enough assumption in lots of situations: one agent, itself rational and evaluative, treats the other agent as if they are non-rational and/or non-evaluative (and hence as if they are acting mechanically, predictably, mindlessly, or habitually). We could then inquire into the mismatch, or fit, of each agent's model with the other agent's model, and hence the extent to which either agent can achieve what it vies for, given the degree of overlap in question. This would be equivalent to A_1 erasing A_2 from its version of figure 9.3, or A_2 erasing A_1 from its version of figure 9.3, such that each agent reduces a relation between relations (the other agent, so to speak) to a mere relation.

While such a scenario is interesting enough in itself, it will be assumed instead that each agent models the other agent's behavior in agentive terms. In particular, A_1 will generate $P_1(S/O)$ using $V_{1(2)}(S/O)$, which is its meta-model of A_2's model. Similarly, A_2 will generate $P_2(I/S)$ using $V_{2(1)}(I/S)$, which is its meta-model of A_1's model. To add depth to this scenario, it will thus be assumed that each agent undertakes, as part of its own evaluative process, its model of the other agent's evaluative process.

In so doing, each agent must turn conditional values into conditional probabilities (e.g., $V_{2(1)}(I/S) \rightarrow P_2(I/S)$ and $V_{1(2)}(S/O) \rightarrow P_1(S/O)$), such that the greater the value of a sign given an object, or of an interpretant given a sign, the greater the probability of the sign given the object, or of the interpretant given the sign. As per the original characterization of relatively rational evaluative processes in chapter 1, it will be assumed that each agent assumes that the other agent strives to maximize the value of its behavior, such that it will always produce the most valuable interpretant given a sign; or always produce the most valuable sign given an object. More precisely, A_1 will assume that A_2 will instigate the most valuable sign (given some object) with probability near 1 (all other signs having a probability near zero). Similarly, A_2 will assume that A_1 will instigate the most valuable interpretant (given some sign) with probability near 1 (all other interpretants having a probability near zero).

As discussed in chapter 1, a useful function for translating values into probabilities with just this pattern is the softmax function, which will be defined and deployed in the following way (omitting, for the moment, the subscripts that indicate agents and embeddings):

$$P(I_i/S_j) = \text{softmax}\big(V(I_i/S_j)\big) = \frac{e^{\beta \cdot V(I_i/S_j)}}{\sum_k e^{\beta \cdot V(I_k/S_j)}}. \tag{9.21}$$

In particular, this function takes parameter β and an array of conditional values $V(I/S)$ as its inputs, and it returns an array of conditional probabilities $P(I/S)$ as its outputs, where the probability of I (given some S) scales with the value of I (given S). Recall that parameter β, where $\beta > 0$, simply determines how contrastive the probabilities are. In particular, as β increases, the probability of the interpretant with the highest value, given some sign, approaches unity, such that the probability of all other interpretants, given the same sign, approaches zero.

While the softmax function has just been exemplified using interpretant-sign relations, it can also be used to translate sign-object relations, such that $P(S_i/O_j) = \text{softmax}\big(V(S_i/O_j)\big)$. In other words, each agent can use the softmax function in its meta-model of the other agent's model. Indeed, as will be shown shortly, they can use it recursively when the embeddings get deep enough.

All that said, it is not necessary to use the softmax function per se: a simple heuristic can also be used. Simply let the probability of the most valuable behavior, given the context, be 1; and the value of all other behaviors, given the context, be zero. We employ the softmax function, in part, to ensure that all the embeddings discussed next involve simple functional application (without reference to an algorithm, *if-then* conditions, and the like).

We will begin with the evaluative perspective of A_2, the signer, and see where the embeddings lead. Suppressing the details of indexical dependency, equation (9.20) becomes

$$V_2(S/O) = \sum_I V_2(I \wedge O) \cdot P_2(I/S) \,. \tag{9.22}$$

If A_2 were to model A_1's behavior as relatively non-agentive, as per the previous discussion, the calculation would stop here. Instead, A_2 models A_1's behavior as agentive, and thus as rational and evaluative. Using the softmax function, as applied to A_2's model of A_1's evaluative process, and thus the recursive version of equation (9.19), the last term in this equation becomes

$$\begin{aligned} P_2(I/S) &= \text{softmax}\left(V_{2(1)}(I/S)\right) \\ &= \text{softmax}\left(\sum_O V_{2(1)}(I \wedge O) \cdot P_{2(1)}(O/S)\right) , \end{aligned} \tag{9.23}$$

which turns on A_2's model of A_1's model. Using Bayes' equation, the last term in this equation becomes

$$P_{2(1)}(O/S) = \frac{P_{2(1)}(S/O) \cdot P_{2(1)}(O)}{\sum_{O'} P_{2(1)}(S/O') \cdot P_{2(1)}(O')} , \tag{9.24}$$

which calculates recursive posteriors using recursive likelihoods and priors. Using the softmax function again, but now as applied to A_2's model of A_1's model of A_2's model, and thus the doubly recursive version of equation (9.20), the likelihood term in this equation becomes

$$\begin{aligned} P_{2(1)}(S/O) &= \text{softmax}\left(V_{2(1(2))}(S/O)\right) \\ &= \text{softmax}\left(\sum_I V_{2(1(2))}(I \wedge O) \cdot P_{2(1(2))}(I/S)\right) . \end{aligned} \tag{9.25}$$

In other words, at this step, A_2 is modeling A_1's model of its own model. If A_2 stopped here, it would be assuming that A_1 was modeling its behavior as non-agentive (as encapsulated in a simple likelihood: $P_{(2(1(2))}(I/S)$). If A_2 assumes, instead, that A_1 is modeling its behavior as agentive, it can go further. In particular, using the softmax function, the last term in this equation becomes

$$P_{2(1(2))}(I/S) = \text{softmax}\big(V_{2(1(2(1)))}(I/S)\big)$$
$$= \text{softmax}\Big(\sum_{O} V_{2(1(2(1)))}(I \wedge O) \cdot P_{2(1(2(1)))}(O/S)\Big)\,, \tag{9.26}$$

which turns on A_2's model of A_1's model of A_2's model of A_1's model. So the analysis could continue, to any level of recursive depth, by doing the mathematical equivalent of clausal embedding, as applied to mental state predicates or speech act verbs:

Albert said that Allison said that Albert said...
I believe that you believe that I believe that...

Rather than continue in this way, perhaps indefinitely, and rather than examine a similar—and perhaps concomitant—recursion that begins with $V_1(I/S)$ as its starting point, we may ask how agents step out of, and otherwise shunt off or route around, such infinite recursions.

First, as mentioned previously, A_2 can simply assume that A_1 is a non-agent, that behaves in a relatively fixed way, corresponding to the likelihood $P_2(I/S)$, such that equation (9.22) is the beginning and end of A_2's evaluative process. Indeed, even if A_2 knows that A_1 is an agent like itself, it can assume that A_1 is a relatively predictable agent, such that the outcome of A_1's evaluative process is more or less equivalent to the likelihood in question. Similar processes often go under the names of *reification*, *naturalization*, *habitualization*, and *normalization*.

Second, and similarly, A_2 can assume that A_1 assumes that A_2 behaves predictably, such that A_2 can model A_1's understanding of its behavior as a simple likelihood. In effect, A_2 locks off at $P_{2(1)}(S/O)$ in equation (9.24), and thus proceeds no further down the chain of embeddings. Crucially, A_2 can use a likelihood that is accurate or not, and intentionally so or unconsciously so. For example, A_2 might assume that A_1 has a good understanding of its behavior, and thus model A_1's likelihood using its own values, such that $P_{2(1)}(S/O) = \text{softmax}\big(V_2(S/O)\big)$. Conversely, A_2 might assume that A_1's model of its behavior is not particularly good, and so model it using a likelihood that does not match its actual behavior. Indeed, A_2 might use the putative inadequacy of A_1's model of its own behavior as a means to dupe A_1, and otherwise secure an advantage.

Third, extending the discussion of agents that learn in Part III, it may be the case that A_1 and A_2 share a code (and perhaps even share the belief that they share the code). Suppose, for example, that A_1 and A_2 need to coordinate their behavior to produce a joint action that benefits them both. It behooves them to signify and interpret using one and the same mapping. In particular, if O_i is mapped to S_j by A_2, then S_j should be mapped to O_i (or I_i) by A_1. Each agent can use this fact when interpreting the behavior of the other.

Fourth, and related to this last point, each of the agents may have told the other how it would behave, and/or have otherwise informed the other what its values and beliefs are, such that each agent already knows the interpretive ground of the other (and each agent knows that both agents know this). Key here is the original definition of a semiotic process: if I bring something to your attention (by pointing, say, such that you turn to look; or simply by stating it, such that you become aware of it), not only does it come to be the case that we are each attending to the same object, it also comes to be the case that we are each attending to the fact that we are each attending to the same object. We can therefore take our joint attention, or intersubjectively held knowledge, for granted as a fundamental presupposition, and use it to guide our subsequent interactions with each other. Crucially, the signs in question do not have to be pointing gestures of the stereotypical sort. They may involve the articulation of beliefs, which include an agent's beliefs about the value and likelihood of certain outcomes, not to mention an agent's beliefs about the nature (or ontology) of the outcomes themselves, and hence knowledge of each of the agent's interpretive grounds—now made more or less explicit, and hence mutually known.

Mutual knowledge, recursive intentionality, conventions, and the creation of intersubjective grounds are obviously large topics, taken up by lots of scholars, such as George Herbert Mead, Erving Goffman, David Lewis, Robert Stalnaker, Paul Grice, Dan Sperber, and Diedre Wilson. So they will not be further explored here. In the next section, rather, we will analyze the relation between intersubjectively held interpretive grounds and culture itself, focusing on signs that indicate the interpretive grounds—qua cultures—of the agents that bear them. As will be seen, it is often precisely such meta-semiotic processes that enable agents to step out of infinite recursivity.

9.5 Culture as Shared Interpretive Grounds

Culture is many things to many people. For present purposes, and in light of this mathematical model of meaning, culture might be best understood as consisting of

- interpretive grounds, along with the semiotic processes that both lead to them and follow from them;
- where such grounds and processes are held in common, relatively speaking, by the semiotic agents that constitute a collectivity;

- where such a commonality is relatively contrastive (insofar as members of other collectivities hold other grounds and processes in common);
- and where members of such collectivities are often reflexively aware, and hence relatively conscious, of their contrastive commonalities.

Such modes of reflexive awareness are typically evinced in meta-semiotic processes. In particular, it may be assumed that members of such collectivities not only recognize themselves (as members of the same collectivity), they also recognize others (as members of different collectivities), by means of meta-signs whose meta-objects are the distinctive interpretive grounds in question. Of interest, then, are the higher-order signs, often borne on the agent's body, and/or embodied in their public practices (such as their semiotic processes), which indicate that the agent's lower-order semiotic processes are mediated by an interpretive ground with a particular structure and content.

Recalling the initial characterization of an interpretive ground as consisting of an ontology, along with a series of commitments (priors, likelihoods, and eventualities) and values, and hence as an ensemble of cognitive and affective propensities, the higher-order objects of such higher-order signs may be represented as follows:

$$\text{meta-}S \xrightarrow[\text{meta-}G]{} \text{meta-}O = G = \begin{cases} O = \{O_1, O_2, \ldots\}, \\ S = \{S_1, S_2, \ldots\}, \\ I = \{I_1, I_2, \ldots\}, \\ C = \{C_1, C_2, \ldots\}, \\ P(O), \\ P(S/O), \\ P(C/I \wedge O), \\ V(C \wedge I \wedge O). \end{cases} \tag{9.27}$$

As may be seen, the relation between such meta-objects and meta-signs is itself mediated by meta-grounds, which may themselves turn on a meta-ontology, as well as meta-commitments and meta-values. These meta-grounds may be recursively represented as follows:

$$\text{meta-}G = \begin{cases} \text{meta-}O = \{\text{meta-}O_1, \text{meta-}O_2, \ldots\} = \{G_1, G_2, \ldots\}, \\ \text{meta-}S = \{\text{meta-}S_1, \text{meta-}S_2, \ldots\}, \\ \text{meta-}I = \{\text{meta-}I_1, \text{meta-}I_2, \ldots\}, \\ \text{meta-}C = \{\text{meta-}C_1, \text{meta-}C_2, \ldots\}, \\ P(\text{meta-}O), \\ P(\text{meta-}S/\text{meta-}O), \\ P(\text{meta-}C/\text{meta-}I \wedge \text{meta-}O), \\ V(\text{meta-}C \wedge \text{meta-}I \wedge \text{meta-}O). \end{cases} \tag{9.28}$$

As may be seen, the meta-objects of such meta-grounds are themselves the interpretive grounds that constitute the generative kernel of that common resource that we called "culture." Not only, then, do such grounds play the role of *dynamic variables* in the model (insofar as they are at once the condition for and consequence of semiotic processes), they also play the role of *fundamental topic* (qua object of discourse) and *foundational resource* (qua valuable good).

But this model of recursive semiotic processes, whereby the objects of higher-order semiotic processes are themselves the grounds of lower-order semiotic processes, is applicable to much more than culture. In particular, any agent may have at its disposal a wide range of interpretive grounds as a function of the context it is in, the action it is undertaking, the role it is performing, the identity of its interlocutor, and other aspects of the situation. Hence, one and the same agent may use one such ground, rather than another, at any given moment when signifying an object or interpreting a sign.

This means that the meta-signs in question need not only indicate that an agent belongs to one collectivity rather than another, they may also indicate that one and the same agent is currently using one interpretive ground rather than another, where such a shift in grounds may be triggered by, or sensitive to, the slightest of contextual cues, qua meta-signs.

Bateson (2000) famously called such meta-signs "keys" and, along with scholars like Goffman (1979), took them to indicate that not only was an agent framing a situation in new terms and/or adjusting its own footing, it might also be—performatively (and so perhaps unsuccessfully)—shifting the framing and footing of their interlocutors, as well as the framing and footing of more distal interactants, as well.

In other words, one agent's switching of grounds could induce, if not coerce, other agents to switch grounds as well, where such switches and swaps of interpretive grounds are, in effect, their meta-interpretants of the other agent's meta-signs, qua context-sensitive rekeyings.

Finally (and hopefully needless to say), there can be meta-meta-signs that make such meta-grounds known, and known to be known, across greater regions of space-time and/or amid larger collectivities of agents. In particular, there may be higher-order signs that lead to relatively shared beliefs regarding how various kinds of agents, in various kinds of contexts, will behave (however error-ridden or ill-willed), insofar as those agents relate to the world via one interpretive ground rather than another, be it as a function of the collectivities they belong to, the beliefs they commit to, or the passions they submit to, not to mention the theories that they have and the prejudices that they hold.

The next section returns to that arch-enemy of shared interpretive grounds, the parasite (and a few of its ever-present companions).

9.6 Enemies, Parasites, and Infrastructure

We now examine a scenario in which there are three agents. Recall part (1) of figure 9.1, which is expanded—and slightly tweaked—as figure 9.4. In this scenario, A_2 can be framed as converting messages (qua objects) into signals (qua signs); and A_3 can be framed as converting signals (qua signs) into messages (qua objects). Crucially, the signals (S') that A_3 receives are not necessarily identical to the signals (S) that A_2 sends, such that the information that is subsequently decoded (as O') is not necessarily the same as the information that was originally encoded (as O). This potential lack of identity occurs because A_1, now understood as a kind of interloper or intermediary, mediates the relation between A_2 and A_3.

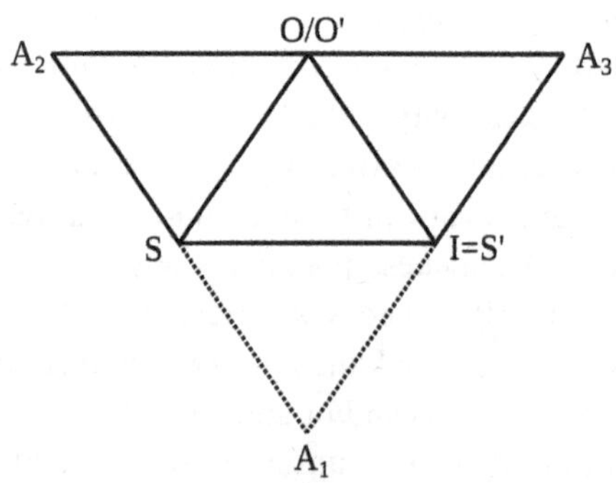

Figure 9.4: Enemies, Parasites, and Infrastructure

This mediating agent (A_1) may be understood, or constituted, in a variety of ways. For example, it may be framed as some kind of channel, environment, or mode of *infrastructure* that carries signals from A_2 to A_3, and thereby bridges some expanse of space or time. Such a channel can carry the signal with greater or lesser degrees of distortion, and hence lesser or greater degrees of fidelity. Shannon's (1948) classic model of a channel, itself more or less subject to noise, could be inserted here. Indeed, we can even model the noisy relation between S and S' as a set of likelihoods, $P(S'/S)$, as was done in section 4.5. In contrast to the usual focus on codes (as mediating between objects and signs), then, we are now focused on channels (as mediating between signifying and interpreting agents, if only in the guise of senders and receivers).

This mediating agent may also be modeled as an *enemy*, and so as an intentional (and malicious) agent rather than a passive conduit. (To be sure, many forms of infrastructure are arguably malicious and active agents in their own right.) In particular, rather than interfering with signals, in the guise of noise, it may intercept signals. It might do this to ensure that no signals get to A_3 (perhaps by generating noise intentionally). Or it might do this to learn the contents of the messages that are sent (and hence as the first part of a decryption process). In short, it can disrupt, intercept, and/or exploit the signals that are sent between the other two agents, thereby disrupting their modes of communication and communion, and hence their attempts to use their interpretive grounds. Shannon's (1949) influential model of secrecy systems, with its emphasis on enemies as opposed to noise, could be inserted here. Recall our analysis of the parasite in chapter 6.

This mediating agent might also be modeled as something like a *translator*, and hence not just as a channel that bridges the distance, be it spatial or temporal, between A_2 and A_3, but also as an interpreting agent that translates the signs that are sent by A_2 into signs that can be understood by A_3. No longer a malicious agent, it might even be understood as a helpful assistant. (That said, there is also a long history of framing translation as treason.)

Finally, any third agent, or an assemblage of such agents, might engage in each of the foregoing modes of mediation, to different degrees, simultaneously and/or sequentially as the signal makes its way across the distance in question—a distance that might even be constituted as a series of interconnected channels, or intermediaries more generally, each with its own additional agent $(A_{1a}, A_{1b}, A_{1c}, \ldots)$.

Crucially, the relation between signifying and interpreting agents (as mediated by physical infrastructure, communicative channels, and social networks) is similar to the relation between objects and signs as well as the relation between signs and interpretants (as mediated by codes, conventions, and interpretive grounds more generally). Indeed, just as the interpretive grounds of agents take into account the relations (between relations) that mediate semiotic processes, such grounds—in an extended sense—also take into account the relations between agents that mediate modes of sociality, however implicitly. Phrased in another way, the values and commitments of agents make reference to, and thereby take into account, not just the components of semiotic processes (such as objects, signs, interpretants, and consequents), and their interrelations, but also the social relations between the signifying and interpreting agents themselves, if only tacitly: doctor and patient, predator and prey, hawk and dove, conspecific and parasite, friend and enemy, novice and expert, translator and traitor, signer and interpreter, and many others. Indeed, in many of the scenarios we analyzed, not only did a shared code come into being, and a relation between agents get established, but the identities and capacities of the agents themselves often emerged in the process. To return to a claim made in chapter 1, interpretive grounds mediate social relations no less than semiotic processes (if such things can be distinguished at all).

But let us leave further exploration of this meta-relation to another book and instead move on to the conclusion.

10 Conclusion

10.1 Synopsis of Arguments

In Part I of this book, the agents had a representation of their environment and its patterns and resources, and they used that representation to make inferences and weigh actions, however erroneously. In Parts II and III of this book, the agents resided in an environment that embodied certain patterns and resources, and they either evolved or learned to act in a way that made sense in the context of that environment, however fleetingly. In the first case, their interpretive ground was mediated by their commitments and values or beliefs and desires. In the second case, their interpretive ground was mediated by the allele frequencies in their genomes, the weights in their registers, or the parameter values in their neurons.

In all such cases, we made sense of their semiotic practices by reference to their interpretive grounds, whether or not their interpretive grounds adequately captured their environments. We likened interpretive grounds to dynamic variables: they are at once shaping of semiotic processes and shaped by semiotic processes. And we saw that environments are, in part, constituted by the semiotic practices and interpretive grounds of other agents, for these too embody patterns and values. Using a mathematical model of the coupling of semiotic processes and interpretive grounds, we thereby analyzed the temporal dynamics of agent-agent and agent-environment relations.

Insofar as the chapters in Parts I, II, and III of this book offered case studies of generalizable scenarios, and insofar as chapter 1, in conjunction with the chapters in Part IV, already offered a synthetic overview of our mathematical model, and its presuppositions and extensions, it would be redundant to review the details here. Rather, this conclusion will briefly take up one final topic: the relation between the interpretive grounds of analysts (who seek to model particular scenarios) and the interpretive grounds of the agents in the scenarios so modeled.

10.2 Revelation and Confrontation

Max Weber, one of the founders of modern sociology, famously introduced the idea of an *ideal type* (2017), understood as a model, or interpretation, of some aspect of social reality. He argued that such models were really representations of unrealities, so they had to be wielded in a somewhat peculiar way if they were to prove useful. In particular, any such model should strive to (1) foreground its own commitments and values, (2) seek evidence of its own inadequacies (as a representation of reality), (3) transform its commitments and values, and otherwise self-update, when confronted with such failures to fit, and (4) be prepared to engage in such a self-reflexive revelation and confrontation again and again, for that is the nature of socially constructed (and hence semiotically mediated) realities, insofar as they are causally coupled to, and thus partially emergent from, our representations of them.

With all this in mind, it might be remarked that our model of interpretive grounds is not only an interpretive ground (in its own terms), but also a Weberian ideal type. In this sense, it constitutes a kind of hermeneutic, if not a potential interpretive paradigm. Framed in another way, this mathematical model is designed to capture the emergence of meaningful processes across a very wide range of agents: not only those that infer and choose, those that evolve, and those that learn, but also (at a meta-level) those that model meaningful processes per se. For we analysts often relate to the outcomes of the systems that we study (which themselves are filled with agents) the way those agents relate to the objects of the signs that they sense. Thus, this book is, in a certain sense, simply one more round in Weber's never-ending circle of revelation and confrontation.

That said, many Weber-inspired scholars, as well as anthropologists, critical theorists, humanists, and pragmatists, will no doubt be wary of the model offered here, insofar as it may seem more focused on logic and mathematics than the raw messiness of actual experience. And so they should be.

But that said, in addition to being merely a model of semiotic processes (and so an ideal type), many of the mathematical techniques deployed here—if only in somewhat souped-up ways—are actually at work in the form of all the algorithmic agents, calculative devices, and computational processes that constitute our world, mediating not only our interactions with each other, but also our relation to the environment and every other meaning-making organism within it. And so, in a certain sense, this is not a model at all, but rather a critical portrait of certain core presumptions of the meaning-making machines that now surround us.

A Energy and Entropy, Information and Value

Imagine an environment in which there are two kinds of thermodynamic systems. Whereas the first kind of system, O_1, has already come to equilibrium with the environment, such that no more useful energy can be extracted from it, the second kind of system, O_2, remains out of equilibrium with the environment, such that useful energy may still be extracted from it. An agent needs to determine which of the two systems it has encountered, so it can decide whether it is worth accepting the system or rejecting it, given the useful energy that it may contain.

Suppose, in particular, that each of the two systems consists of N particles, distributed across two energy levels, E_1 and E_2, in contact with a heat reservoir at temperature T (see figure A.1). In the at-equilibrium system, the barrier between the two levels, as constituted by some activation energy, is very small, so the particles can move back and forth freely. We know from statistical mechanics that the system follows a Boltzmann distribution, such that the probability of finding a particle in a particular energy level, E_i, is proportional to $e^{-E_i/kT}$ (where k is Boltzmann's constant). In the out-of-equilibrium system, in contrast, the activation energy is very large. So there is effectively a barrier between the two energy levels, such that if there are initially n_1 particles in the E_1 state and $n_2 = N - n_1$ particles in the E_2 state, the system will stay that way until the barrier is removed. In particular, the probability of finding a particle in a particular state is simply n_i/N, regardless of the temperature. The barrier is thus akin to a dam that is storing energy (and/or preserving order). If the agent finds such an out-of-equilibrium system, it can remove the constraint and direct the energy of the system to some useful end.

In other words, O_2 systems store free energy and, when coupled with other systems (and potentially catalyzed accordingly, such that the constraint is lowered or removed), they can cause otherwise non-spontaneous processes to occur. In being more out of equilibrium, and hence "livelier" than their O_1 counterparts, such systems are thus more valuable to an agent, insofar as the agent needs free energy to do work.

Imagine that a semiotic agent inhabits an environment in which both kinds of objects, as thermodynamic systems, are equally likely to be found. When the agent encounters any such system, it performs an assayal by examining one of the system's N particles at random to see whether that particle is in the low-energy state ($S_1 = E_1$) or the high-energy state ($S_2 = E_2$). From this assayal, it infers whether the system is at-equilibrium (O_1), or out-of-equilibrium (O_2). And, having made such an inference, it decides whether to accept (I_1) or reject (I_2) the system. If it accepts the system, it pays

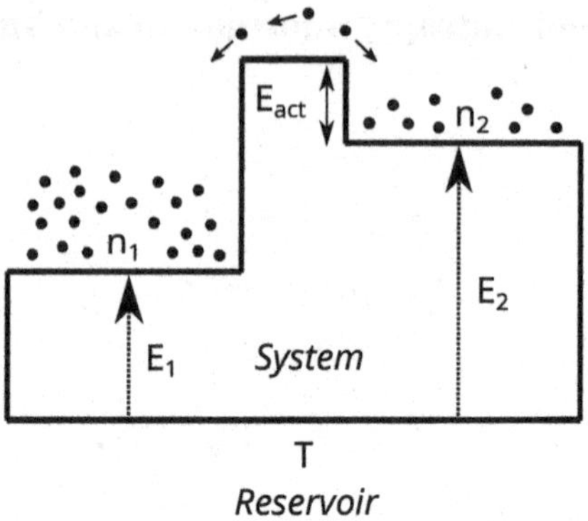

Figure A.1: A Simple Thermodynamic System

a cost equal to the average value of the two systems and obtains whatever value the system actually has, where a system's value is proportional to the free energy that it contains. If it rejects the system, it pays nothing and obtains nothing. Loosely speaking, the ontology, commitments, and values of the agent in this scenario are as closely coupled to 'nature' as can be. We are interested in determining both the surprise value (or information) and exchange value (or price) of such signs.

In a certain sense, then, this scenario is a radical extension, and naturalization, of Laplace's urn. It is not meant to be particularly realistic. Rather, it is meant to tightly couple thermodynamic analysis with semiotic analysis, Shannon entropy with Boltzmann entropy, value with free energy, and gradients with meaning. In what follows, we offer a brief mathematical exploration of its most relevant entailments.

Both systems are equally likely to be found in the environment, which means that the priors are flat:

$$P(O_1) = 1 - P(O_2) = 1/2\,. \tag{A.1}$$

The probability that a particle is in a particular energy state in the at-equilibrium system, which is equal to the likelihood that it exhibits a particular sign, is given by Boltzmann's distribution:

$$P(S_i/O_1) = P(E_i/O_1) = \frac{e^{-E_i/kT}}{e^{-E_1/kT} + e^{-E_2/kT}}\,. \tag{A.2}$$

Stated more carefully, the lower the energy level, the more likely a particle will be found there; as the temperature decreases, all the particles will be found in the low-energy state; and as the temperature increases, the particles will be equally likely to be found in either state. Figure A.2 plots this distribution as a function of T for two arbitrarily chosen energy levels.

(Note that we have set $k = 1$, so are measuring energy in units equal to 1.38×10^{-23} of a joule. Temperature is being measured in kelvins. Finally, entropy is being measured in such energy units per kelvin.)

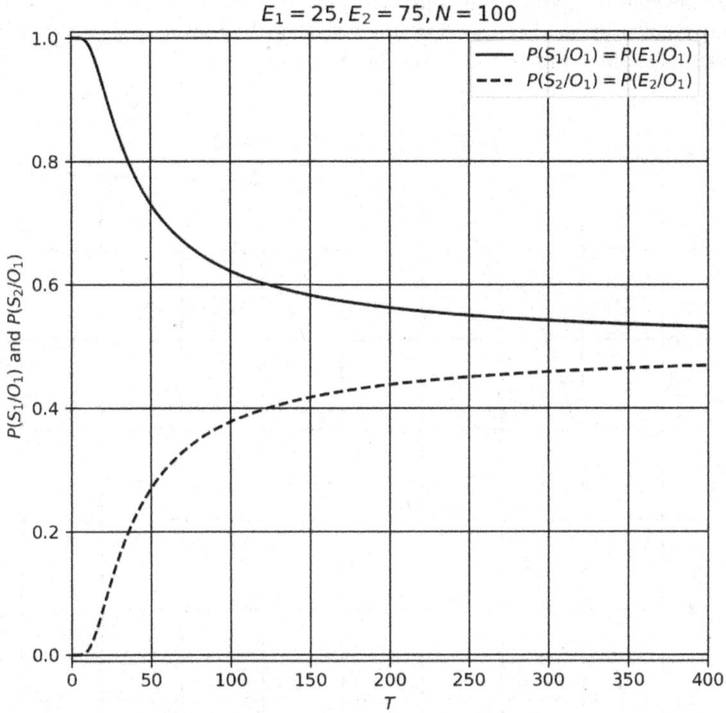

Figure A.2: The Boltzmann Distribution

The probability that a particle is in a particular energy state in the out-of-equilibrium system, which is equal to the likelihood that it exhibits a particular sign, is given by the initial conditions of that system:

$$P(S_i/O_2) = P(E_i/O_2) = \frac{n_i}{N} . \tag{A.3}$$

These will be parameterized, for later adjustment and comparison, using the variable r, such that $n_1/N = r$ and $n_2/N = 1 - r$. In other words, if $r = 1$, all the particles are in the low-energy state; and if $r = 0$, all the particles are in the high-energy state.

These likelihoods may be used to find the Boltzmann entropy of each system:

$$H(O_i) = -kN \sum_j P(S_j/O_i) \cdot \ln P(S_j/O_i) . \tag{A.4}$$

Given the fact that its likelihoods are governed by the Boltzmann distribution, the entropy of the at-equilibrium system, $H(O_1)$, has the following characteristics: The lower the temperature, the more the particles are likely to be found in the low-energy state, so the entropy goes towards zero. The greater the temperature, the more the particles are equally likely to be found in either state, so the entropy goes towards $N\,k\ln 2$ (its maximum value). Figure A.3 shows this behavior.

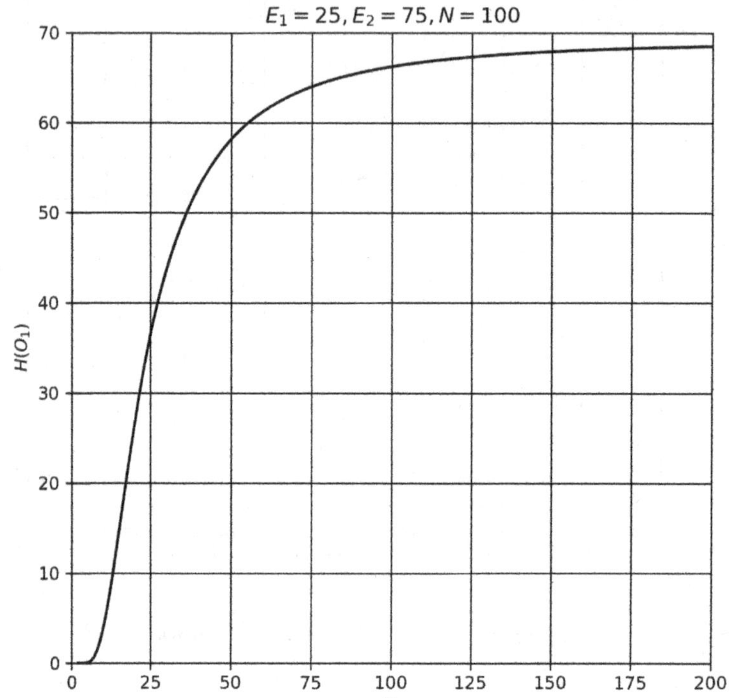

Figure A.3: Entropy of the At-Equilibrium System

Note that Boltzmann entropy differs from Shannon entropy, in that it turns on the natural logarithm (involving base e rather than base 2) and it is multiplied by Boltzmann's constant k. Removing those two differences, however, it can still be used as a measure of our uncertainty regarding which state the system is in. At low temperatures, we are maximally certain of the system's configuration, for all the particles are in the low-energy state; at high temperatures, we are maximally uncertain of the system's configuration for all the particles are equally likely to be in either state.

The likelihoods may also be used to find the expected energy of each system:

$$E(O_i) = N \sum_j E_j \cdot P(S_j/O_i)\,. \tag{A.5}$$

And the results of these last two calculations may be used to find the Helmholtz free energy of each system:

$$F(O_i) = E(O_i) - T \cdot H(O_i) \,. \tag{A.6}$$

As may be seen, the free energy is the energy of the system minus the entropy of the system (times the temperature). It represents the available energy of the system, and hence that part of the system's energy that can be used to do work. Crucially, what it means for a system to be at equilibrium is that its free energy attains its minimal value (which is to say that the entropy of the system, plus the reservoir, attains its maximum value). The free energy of an out-of-equilibrium system is thus greater than the free energy of a system at equilibrium.

Figure A.4 plots the average energy and free energy of the at-equilibrium system (O_1) as a function of temperature. As before, $k=1$, $N=100$, $E_1=25$ and $E_2=75$. As may be seen, at low temperatures all the particles are in the lower-energy level, so the energy is just N times that level. At high temperatures, in contrast, the particles are evenly distributed between both levels, so the energy is just N times the average energy of the two levels. As the difference between the energy and the product of the temperature and entropy, the free energy decreases as the temperature increases, reaching zero when $E(O_1) = T \cdot H(O_1)$. In effect, the higher the temperature, the more the energy of the at-equilibrium system takes the form of relatively low-grade, disorganized thermal energy.

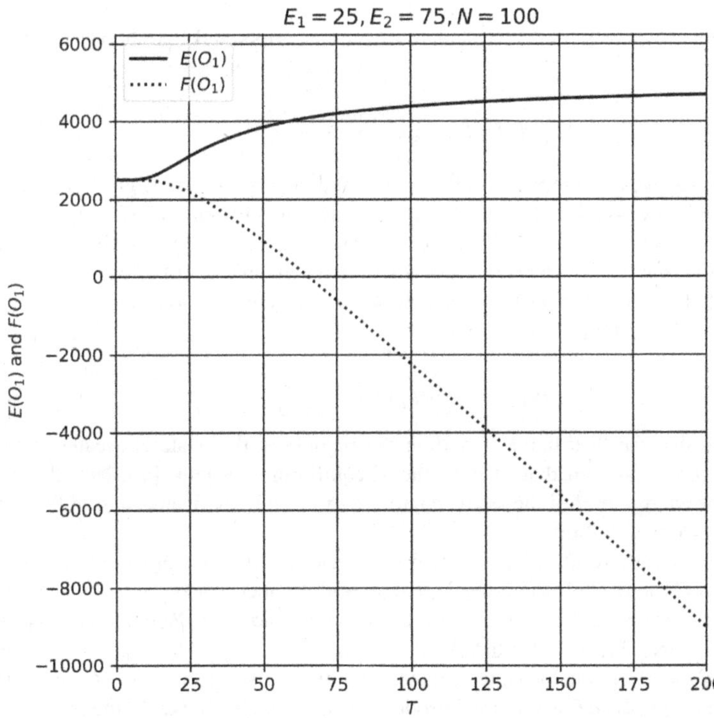

Figure A.4: Energy and Free Energy of the At-Equilibrium System

So much for the thermodynamics of each system—let us now return to the economics and semiotics of the scenario. We assume that the real (exchange) value of a system is the difference between its free energy and the free energy of a system in equilibrium. In other words, $V(O_1)=F(O_1)-F(O_1)=0$ and $V(O_2)=F(O_2)-F(O_1)$. Stated more carefully, the real value of a system is equal to the difference between its free energy and the free energy of an otherwise identical system that has already come to equilibrium. We assume that it costs the average value of both systems, $\overline{V}=\left(F(O_2)-F(O_1)\right)/2$, to accept a system. And we assume that the agent receives the value of a system if it is accepted, and nothing if it is rejected. The payoff matrix is thus:

$$V(I\wedge O)=\begin{bmatrix} V(O_1)-\overline{V} & V(O_2)-\overline{V} \\ 0 & 0 \end{bmatrix} \tag{A.7a}$$

$$=\begin{bmatrix} \frac{F(O_1)-F(O_2)}{2} & \frac{F(O_2)-F(O_1)}{2} \\ 0 & 0 \end{bmatrix}. \tag{A.7b}$$

The conditional values may be found using the reduced version of our main equation from chapter 1 (insofar as we are ignoring consequents):

$$V(I_i/S_j)=\sum_k V(I_i\wedge O_k)\cdot P(O_k/S_j), \tag{A.8}$$

where the posteriors may be calculated from the priors and likelihoods using Bayes' theorem:

$$P(O_j/S_i)=\frac{P(S_i/O_j)\cdot P(O_j)}{\sum_m P(S_i/O_m)\cdot P(O_m)}. \tag{A.9}$$

As usual, we are interested in the agent's judgments. Given a sign, $S_i=E_i$, an agent needs to know whether it better to accept a system (insofar as $V(I_1/S_i)>V(I_2/S_i)$) or reject the system (insofar as $V(I_1/S_i)<V(I_2/S_i)$). Doing the algebra, the first result we find is that the agent's decision depends on Boltzmann's distribution. In particular, if the particle is found in energy level E_i when the system is assayed, then it is worth accepting the system, so long as

$$P(E_i/O_2)>P(E_i/O_1)=\frac{e^{-E_i/kT}}{e^{-E_1/kT}+e^{-E_2/kT}}, \tag{A.10}$$

meaning that the likelihood of finding the particle in that state is greater in the out-of-equilibrium system than it is in the at-equilibrium system. In effect, the agent is using the energy level of the assayed particle as a highly motivated sign of the system's distance from equilibrium.

Such a simple result arises for two main reasons: how we assessed the value of a system and the cost of accepting a system; and the fact that both systems are equally likely to be found in the environment. Under such conditions, Boltzmann's distribution sets the critical threshold for the choice scenario.

Using the foregoing ideas, we can now determine the information of a sign, and the value of a sign. As defined in the introduction and used in chapter 2, the information of a sign is the average number of yes-or-no questions that it saves the agent from having to ask in order to learn the identity of the system:

$$\mathcal{I}(S)=\sum_i P(S_i)\cdot\left[H(O)-H(O/S_i)\right], \tag{A.11}$$

where the entropy and conditional entropy are as follows:

$$H(O)=-\sum_j P(O_j)\cdot\log_2 P(O_j)=1, \tag{A.12a}$$

$$H(O/S_i)=-\sum_j P(O_j/S_i)\cdot\log_2 P(O_j/S_i). \tag{A.12b}$$

Note that we are now using Shannon entropy to calculate our uncertainty as to which system we have (at-equilibrium or out-of-equilibrium). Earlier, we used Boltzmann entropy to calculate our uncertainty as to the particular arrangement of particles in a system, and hence as a measure of its disorder. The two notions and their applications are closely related, but not identical.

Finally, we may calculate the value of a sign:

$$\begin{aligned}\mathcal{V}(S)&=\sum_i P(S_i)\cdot\mathcal{V}(S_i)\\&=\sum_i P(S_i)\cdot\left[\max\left(V(I/S_i)\right)-\max\left(V(I)\right)\right],\end{aligned} \tag{A.13}$$

where $V(I/S)$ is given by equation (A.8) and $V(I)$, understood as the value of an interpretant without a sign, is given by the same equation with the posteriors replaced by priors:

$$V(I_i)=\sum_j V(I_i\wedge O_j)\cdot P(O_j). \tag{A.14}$$

Figure A.5 shows the information of a sign, as a function of r and T, where $E_1=25$ and $E_2=75$. As may be seen, for each value of T, there is a value of r where the information of a sign is zero. This is because at that temperature the two systems are identical, insofar as the $P(E_i/O_1)=P(E_i/O_2)$. In other words, the out-of-equilibrium system happens to have the same distribution as a Boltzmann distribution. Moving away from this minimum in either direction (and hence toward higher values of r or lower values of r), the information of a sign increases. As the temperature goes to zero, all particles in the at-equilibrium system will be found in the low-energy state, so signs become more informative the more the out-of-equilibrium system has particles in the high-energy state ($r\rightarrow 0$). Indeed, signs are maximally informative, meaning that after assaying the system, the agent is completely certain of its identity, if $T=0$ and $r=0$: in the at-equilibrium system, all the particles are in the low-energy state, and in the out-of-equilibrium system, all the particles are in the high-energy state. As may also be seen, for low values of r, the information of a sign decreases as the temperature increases. Whereas for high values of r, the information of a sign increases as the temperature increases. Finally, as should be expected, the information of a sign is always between 0 and 1 bits, as the sign can only help distinguish between the two types of systems.

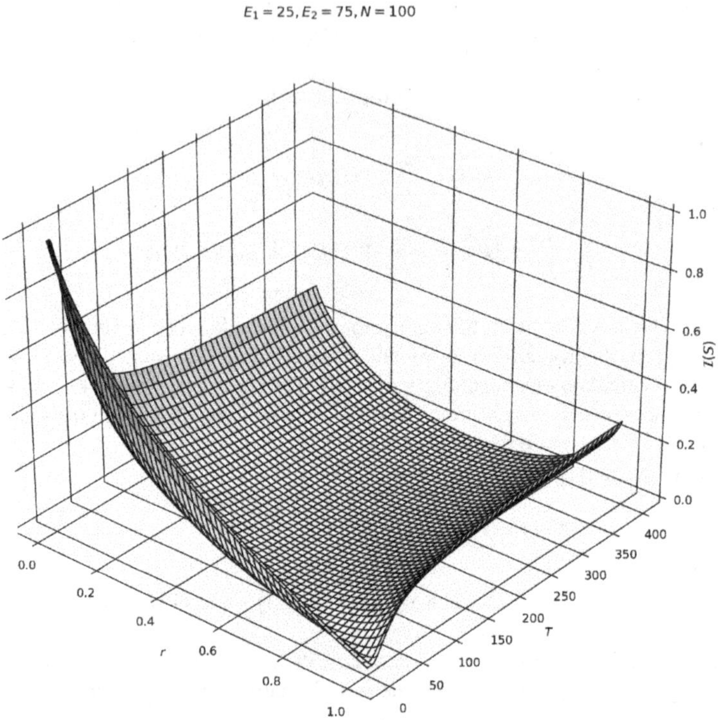

Figure A.5: The Information of a Sign

Figure A.6 shows the value of a sign for systems with the same conditions: how much an agent would be willing to sacrifice for a sign (in the currency of free energy), insofar as that sign would help the agent make a decision regarding whether to accept a system (and thereby acquire the free energy that it contains). As with information, the value of a sign reaches a minimum wherever the two systems are identical. Again, then, the bottom of the valley traces a Boltzmann distribution: values of r and T at which the agent is undecided between the two systems. And the value increases, for a given value of T, as r moves away from its equilibrium value (again, in either direction). As may also be seen, all else being equal, a sign is more valuable for values of r near zero than for values of r near 1. This is because the particles in the out-of-equilibrium system are in the high-energy state rather than the low-energy state, so the system is storing energy (in addition to order). Finally, the value of the sign tends to increase as T increases. This is because the free energy of the at-equilibrium system gets smaller and smaller, while the free energy of the out-of-equilibrium system stays constant (insofar as it is independent of T).

Note, then, in comparing figure A.5 and figure A.6, that the value of a sign can be increasing while the information of a sign may be decreasing (as we move through

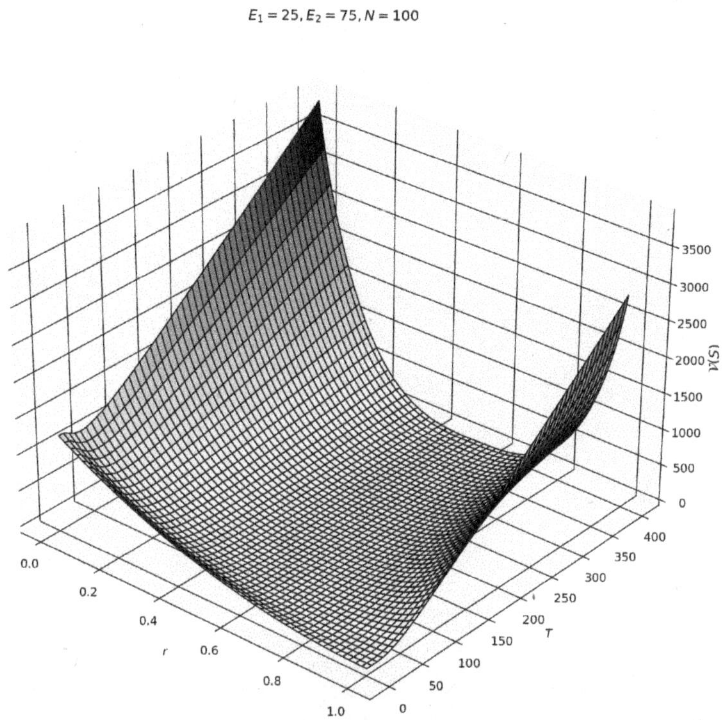

Figure A.6: The Value of a Sign

parameter space). This is especially true for small values of *r*, and hence when the particles in the out-of-equilibrium system are in the high-energy state.

In short, attending to the energy level of a randomly assayed particle provides a highly informative and valuable sign of a system's configuration, telling the agent whether the system remains out-of-equilibrium, and hence whether it is worth sacrificing energy to obtain that system.

B Complexity, Organization, and Constraint

While entropy, in the guise of Shannon information, is not the best measure of meaning, it is a useful measure of complexity, organization, and constraint, especially when it is generalized to scope over all the constituents in an agent's ontology. This is especially true when we take the view of the analyst (of the agent's behavior), rather than the view of the agent per se.

From chapter 8, recall that an agent's ontology—insofar as it consists of objects, signs, interpretants and consequents—determines a *universe*, understood as a relevant set of possible worlds, where each such world is a unique way that the universe might unfold, given the ontology in question.

In the symptom and sickness scenario analyzed in chapter 2, there were two objects, two signs, two interpretants, and two consequents (understood as independently specifiable events). We will let w_{ijkl} denote the world in which the object that obtains is O_i, the sign that obtains is S_j, the interpretant that obtains is I_k, and the consequent that obtains is C_l. For example, w_{1222} is the world in which the object is O_1 (person was healthy), the sign is S_2 (person was symptomatic), the interpretant is I_2 (person took medicine), and the consequent is C_2 (person became ill). The universe for this scenario, denoted as Ω, therefore has the following 16 worlds:

$$\begin{aligned}\Omega = \{ & w_{1111}, w_{1112}, w_{1121}, w_{1122},\\ & w_{1211}, w_{1212}, w_{1221}, w_{1222},\\ & w_{2111}, w_{2112}, w_{2121}, w_{2122},\\ & w_{2211}, w_{2212}, w_{2221}, w_{2222} \}.\end{aligned} \tag{B.1}$$

Rather than use Shannon entropy to calculate the information in a sign, as was done in chapter 2, we will use it to calculate the complexity of this universe: where the more complex a system is, the less certain we are—as analysts—as to the state it is in. To do this, we need to know the probability of each world within the universe: $P(w_{ijkl})$. This is equivalent to finding the joint probability distribution of all the worlds delimited by the ontology. As derived in chapter 8, such a distribution may be calculated using the priors, likelihoods, eventualities, and values that constitute the interpretive ground:

$$\begin{aligned}P(w_{ijkl}) &= P(O_i \wedge S_j \wedge I_k \wedge C_l)\\ &= P(C_l/I_k \wedge O_i) \cdot P(I_k/S_j) \cdot P(S_j/O_i) \cdot P(O_i).\end{aligned} \tag{B.2}$$

In contrast to Shannon's notion of information, then, we are not interested in the probability of a sign (given a set of signs, or code). We are interested, rather, in the probability of a world (given a set of worlds, or universe). And, in contrast to relative entropy, or Kullback–Leibler divergence, we are not interested in the constraints imposed on an object by the existence of a sign. We are interested, rather, in the constraints imposed on a universe (which may include the presence of agents with particular interpretive grounds), insofar as such constraints make certain worlds more or less likely, and thereby make the universe less complex (insofar as it becomes more patterned, and thus more predictable).

Using these probabilities, the Shannon entropy of the universe (expressed in bits) is simply the average improbability of all the worlds within the universe:

$$\begin{aligned} H(\Omega) &= \sum_{w \in \Omega} P(w) \cdot \log_2 \frac{1}{P(w)} \\ &= \sum_{ijkl} P(w_{ijkl}) \cdot \log_2 \frac{1}{P(w_{ijkl})} . \end{aligned} \tag{B.3}$$

Framed in the language of uncertainty, this measure gives us the number of yes-or-no questions that we would have to ask to determine which state the universe is in, and hence which world obtains. Framed in the language of agency, this is the number of binary decisions we would have to make to put the universe into a particular (and fully determined) state. This is what we meant when we said that we are now taking the perspective of an analyst of the scenario rather than that of an agent in the scenario.

While complexity, in the guise of Shannon entropy, is an interesting notion, it is often more useful to measure the *organization* of a universe. This is the difference between the maximum entropy of a universe (when there are no constraints on a universe such that each world within it is equally likely) and the actual entropy of the universe (when constraints have been imposed such that some worlds in the universe are more probable than others). Organization may thus be measured as follows:

$$\mathcal{O}(\Omega) = H_{max}(\Omega) - H_{act}(\Omega) . \tag{B.4}$$

As used here, *constraints* are simply all the biases, correlations, patterns, causal relations, laws, propensities, and agentive strategies that order a universe, such that certain worlds (qua possible ways that the universe might unfold) are more or less likely than others. In the scenario analyzed in chapter 2, involving symptoms and sickness, such constraints were modeled by p, a, b, q, r, and c, the parameters underlying the priors, likelihoods, eventualities, and values of the agent. Depending on the values of such parameters, and hence the relative probability of each possible world, a universe may be more or less organized.

Insofar as $|\Omega|$ denotes the cardinality of the universe, and hence the number of worlds in that universe (understood as a set of elements), the probability of each world in the universe when there are no constraints (nc) is simply

$$P_{nc}(w_{ijkl}) = \frac{1}{|\Omega|} = \frac{1}{16} . \tag{B.5}$$

We can use these non-constrained probabilities, in conjunction with equation (B.3), to calculate the maximum entropy of the universe:

$$
\begin{aligned}
H_{max}(\Omega) &= \sum_{w \in \Omega} P_{nc}(w) \cdot \log_2 \frac{1}{P_{nc}(w)} \\
&= \sum_{ijkl} P_{nc}(w_{ijkl}) \cdot \log_2 \frac{1}{P_{nc}(w_{ijkl})} \\
&= \sum_{ijkl} \frac{1}{|\Omega|} \cdot \log_2 |\Omega| \\
&= |\Omega| \cdot \frac{1}{|\Omega|} \cdot \log_2 |\Omega| \\
&= \log_2 |\Omega| \\
&= \log_2 16 \\
&= 4 \text{ (bits)}.
\end{aligned} \tag{B.6}
$$

Given the characterization of (Shannon) entropy offered here, this answer should come as no surprise. Assuming that each world is equally likely, we only need to ask four yes-or-no questions to determine which of those possible worlds is the actual world: the first question determines whether O_1 or O_2 obtains; the second question determines whether S_1 or S_2 obtains; the third question determines whether I_1 or I_2 obtains; and the fourth question determines whether C_1 or C_2 obtains. Universes with other ontologies may have more or fewer worlds, and hence require more or fewer questions to arrive at such an answer.

Having calculated the maximum entropy, the organization of the universe—and, in particular, how the organization varies as a function of various parameter values—is readily calculated:

$$
\begin{aligned}
\mathcal{O}(\Omega) &= H_{max}(\Omega) - H_{act}(\Omega) \\
&= \sum_{w \in \Omega} P_{nc}(w) \cdot \log \frac{1}{P_{nc}(w)} - \sum_{w \in \Omega} P_c(w) \cdot \log \frac{1}{P_c(w)} \\
&= \log_2 |\Omega| - \sum_{ijkl} P_c(w_{ijkl}) \cdot \log \frac{1}{P_c(w_{ijkl})},
\end{aligned} \tag{B.7}
$$

where $P_c(w_{ijkl})$ denotes the probability of each world in the universe when constraints (c) are imposed. As expressed in equation (B.2), and now properly rechristened, it is simply

$$
P_c(w_{ijkl}) = P(C_l/I_k \wedge O_i) \cdot P(I_k/S_j) \cdot P(S_j/O_i) \cdot P(O_i), \tag{B.8}
$$

where each of the terms in this equation was given in chapter 2 as a function of the parameter values.

Note, then, that minimal organization is the same as maximal actual entropy (for the ontology, and/or interpretive ground, in question). Stated more generally, the less constrained a universe is, given the parameters of the scenario, the less organized it is, and hence the greater our ignorance (or the less our certainty) as to the particular world that obtains. Alternatively, the more constrained a universe is, the less complex it is and/or the more organized it is, and hence the greater our certainty as to the particular world that obtains.

Indeed, if the universe is so organized that the probability of some particular world is 1, and the probability of all other worlds is zero, then the actual entropy of the universe is zero, such that the organization of the universe is equal to its maximum entropy. In effect, the constraints governing the universe are so strong that nothing new or surprising ever happens there.

There is a term in equation (B.8), $P(I_k/S_j)$, which is as yet undefined. In effect, it is the probability that the agent instigates interpretant I_k conditioned on sensing sign S_j. As discussed in section 2.3 of chapter 2, assuming that the agent acts according to the values it holds, such probabilities may be generated from the values $V(I_k/S_j)$ that were calculated using the main equation from chapter 1. In particular, if the agent instigates the most valuable interpretant it can, given the sign it senses, these conditional probabilities are

$$P(I_1/S_j) = 1 - P(I_2/S_j) = \begin{cases} 1 & V(I_1/S_j) > V(I_2/S_j)\,, \\ 0 & V(I_1/S_j) < V(I_2/S_j)\,, \\ \frac{1}{2} & \text{if } V(I_1/S_j) = V(I_2/S_j)\,. \end{cases} \tag{B.9}$$

In the plots that follow, rather than use such conditional probabilities as such, we will employ the usual softmax function: $P(I/S) = \text{softmax}\big(V(I/S)\big)$. So long as parameter β is large, this gives nearly identical results, with the added benefit of being a continuous function.

Figure B.1 plots the organization of this universe as a function of p for various values of a and b, holding the other parameter values constant. To understand the shape of these curves, focus first on the solid line, which plots $\mathcal{O}(\Omega)$ versus p when $a = b = .01$. As may be seen, the organization is highest when $p = 1$, relatively high when $p = 0$, and lowest when $p \approx .35$. When $p = 1$, the priors are such that $P(O_1) = 1$, so we know for certain that the person is healthy. This rules out 8 of the 16 possible worlds immediately—in particular, all worlds w_{ijkl} in which $i \neq 1$. Insofar as a and b are very small, signs are strongly correlated with objects; and insofar as the person is healthy, they are very likely to be asymptomatic, such that $P(S_1) \approx 1$. This effectively rules out four more worlds—in particular, all the remaining worlds w_{1jkl} in which $j \neq 1$. From this first set of factors alone (high p and low a and b), the actual complexity of the universe is already half of its maximum complexity, thereby contributing two bits to the overall organization. Given the fact that the person, as a semiotic agent, is very certain that they are healthy (given the sign that they received, along with their original priors), they almost certainly do not take the medicine, such that $P(I_1) = 1$. This rules out two more worlds (those remaining worlds w_{11kl} in which $k \neq 1$), and thereby adds one more bit to the organization of the universe. Finally, as may be seen by the eventualities, $P(C_1/I_1 \wedge O_1) = 1$, such that a healthy person (O_1) who does not take the medicine (I_1) is certain to remain healthy (C_1). This rules out one of the last two worlds, such that only w_{1111} remains. This adds another bit to the organization of the universe, such that $\mathcal{O}(\Omega) \approx 4$ bits.

Continuing to focus on the solid line, a very similar set of reasons motivates the fact that the organization of the universe is also relatively high when $p = 0$. In particular, the object that obtains is certainly O_2 (person is sick); and, given how small a and b are, the sign that obtains is very likely S_2 (person is symptomatic). Given this information,

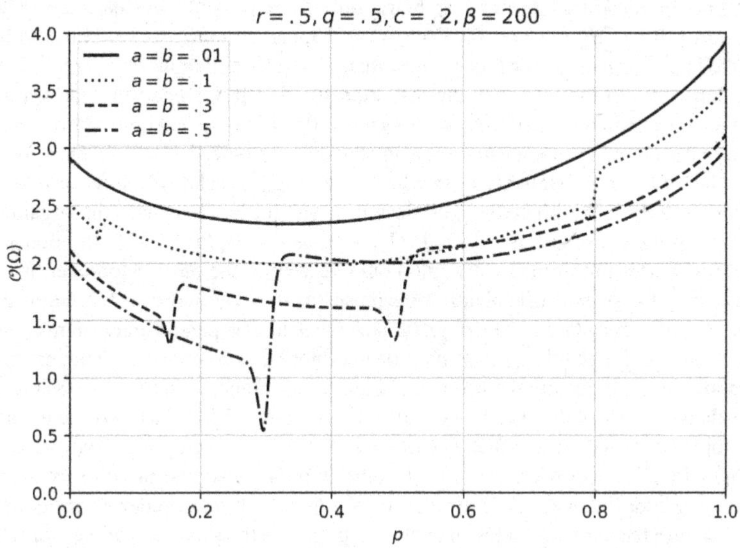

Figure B.1: Organization of the Universe as a Function of p

the person is very likely to instigate I_2 (and thereby take the medicine). In effect, then, all possible worlds have been ruled out except w_{2221} and w_{2222}. Finally, as may be seen by the eventualities, as well as the fact that $r = .5$ (meaning that the medicine is not completely efficacious), it is not certain that the person will get better. Indeed, these two remaining possible worlds are almost equally likely: $P(w_{2221}) \approx P(w_{2222}) \approx 1/2$. The organization of the universe, therefore, never gets above 3 bits.

It is now easy to understand why the organization of the solid line reaches its minimum around $p = .35$. In part, this is because the person may or may not be sick: $P(O_1) = 1 - P(O_2) = .35$. This means that the person may or may not be symptomatic: $P(S_1) = 1 - P(S_2) \approx .35$. In other words, it is no longer easy to rule out possible worlds just by knowing the value of p alone. That said, given the fact that a and b are so small, we know that O_1 worlds are almost certainly S_1 worlds and O_2 worlds are almost certainly S_2 worlds. And this kind of correlation is precisely a constraint. If we look at the person's conditional values, $V(I/S)$, when $p = .35$, we see that their behavior is completely determined by the sign they receive. If the person is asymptomatic, they will definitely not take the medicine, insofar as $V(I_1/S_1) > V(I_2/S_1)$, such that $P(I_1/S_1) = 1$; and if the person is symptomatic, they will definitely take the medicine, insofar as $V(I_2/S_2) > V(I_1/S_2)$, such that $P(I_2/S_2) = 1$. In other words, O_1 and S_1 worlds are almost definitely I_1 worlds and O_2 and S_2 worlds are almost definitely I_2 worlds. Given the eventualities of this scenario, if the person is healthy and does not take the medicine, they definitely remain healthy. This means that O_1, S_1, and I_1 worlds are definitely C_1 worlds. However, if the person is sick (and so is definitely symptomatic, and so definitely takes the medicine), they are just as likely to get better as they are to get sick, insofar as the medicine is not particularly efficacious ($r = .5$). In short, given all the constraints in question, there end up being only three possible worlds, each of which is similarly likely: $P(w_{1111}) \approx .34$, $P(w_{2221}) \approx .33$, and $P(w_{2222}) \approx .33$. Thus, while the universe is less organized for mid-range values of p, it is still quite organized, given the particular values of the other parameters, understood as pattern-causing constraints.

Let us move from the solid line, where $a = b = .01$, to the dashed line, where $a = b = .3$. As may be seen, it is somewhat similar in shape to the other lines: it reaches its maximum when $p = 1$; it is relatively high when $p = 0$; and it is relatively low for mid-range values of p. Given the preceding discussion, it should be clear why the organization is so high at the extremes: as $p \to 1$, $P(O_1) \to 1$; as $p \to 0$, $P(O_2) \to 1$. In other words, half of the worlds are immediately ruled out one way or the other. Moreover, it should be clear why the overall organization decreases as $a \to 1/2$ and $b \to 1/2$. Signs are no longer strongly correlated with objects, such that a sick person may or may not be symptomatic and a healthy person may or may not be asymptomatic. This means that one cannot rule out, just by the priors and likelihoods alone, particular possible worlds.

All that said, we still need to explain why the dashed line has two local minima. If we look at $P(I/S)$, we see that $P(I_1/S_2) = P(I_2/S_2) \approx .5$ when $p = .5$; and we see that $P(I_1/S_1) = P(I_2/S_1) \approx .5$ when $p = .155$. In other words, these two minima are precisely the points where the agent is undecided regarding which interpretant to instigate; and this is because the expected value of each interpretant is the same. Recall equation (B.9), and the analysis of critical prices in chapter 2. This indecisiveness on the agent's part means that its action is minimally predictable at such points (at least when the agent senses the sign in question). This means that the consequents are similarly unpredictable. And all this means that the organization of the universe, at precisely such critical points, reaches its minimum values.

Finally, a reader may wonder why the organization is higher immediately to the right of such minima, and lower immediately to the left. This is because, immediately to the right of such minima, the agent again becomes decisive, instigating I_1, whereas immediately to the left of such minima, the agent again becomes decisive, instigating I_2. Given the eventualities of this scenario, instigating I_1 (no medicine is taken) does not affect the person's health: if they were sick, they remain sick; and if they were healthy, they stay healthy. In contrast, instigating I_2 (medicine is taken), can affect the person's health, for better or for worse. In short, the action undertaken to the immediate right of a minimum (I_1) has a more predictable effect than the action undertaken to the immediate left of a minimum (I_2). Similar arguments and observations motivate the shape of the other two lines in this graph.

To give readers another perspective on the relation between complexity, constraints, and agency, figure B.2 plots the organization of the universe as a function of $a = b$ and p (while holding the values of r, q, and c constant).

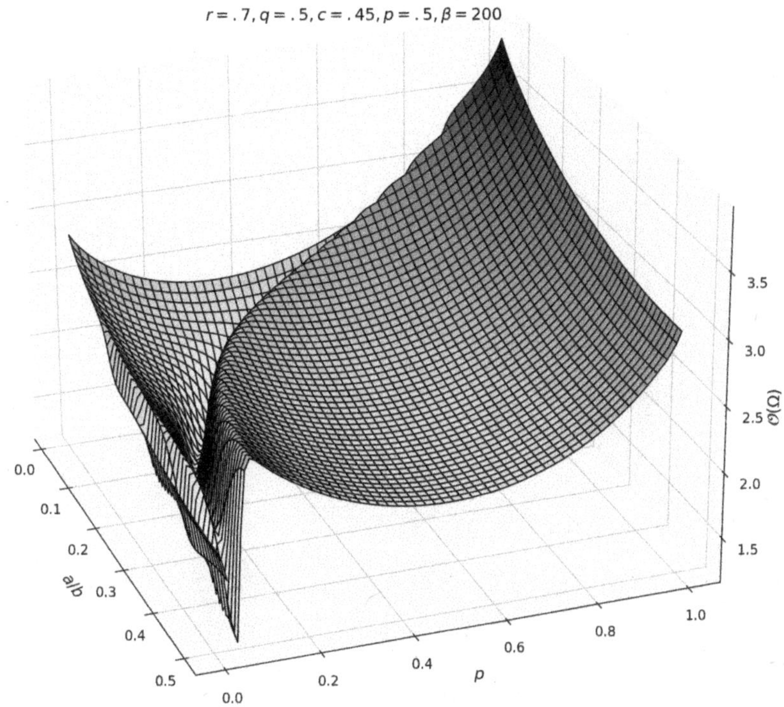

Figure B.2: Valley of Indecision

Broadly, and as should be expected given the preceding discussion, the organization of the universe is higher when p approaches extreme values and when the values of a and b are relatively small. As may also be seen, there is a deep valley where the two-dimensional surface is sharply folded and the organization of the universe reaches its minimum value. Given the foregoing discussion, it should be clear that the values of $a = b$ and p that trace out the very bottom of this valley are precisely those points where the agent is undecided regarding which interpretant to instigate (insofar as the expected values of I_1 and I_2, for a given sign, are the same).

C The Utility of a Sign

It is worth remembering that the main equation (from chapter 1) turns on expected value as opposed to expected utility. For readers interested in microeconomic approaches to meaning, a simple way to incorporate utility into the analysis is to replace $V(C \wedge I \wedge O)$ with $U\big(V(C \wedge I \wedge O)\big)$ in the main equation, where

$$U(x) = (x + M)^{\alpha} . \tag{C.1}$$

Parameter M denotes the money that an agent has in all worlds, regardless of the outcome of the scenario being modeled. In effect, it is the money—or value more generally—that they have in the bank, as opposed to the money that they stand to gain or lose in the scenario itself. Parameter α, in contrast, characterizes the degree to which the agent is risk-sensitive (where $0 < \alpha < 1$). In particular, as $\alpha \rightarrow 0$, the agent being modeled becomes more and more risk-adverse; conversely, as $\alpha \rightarrow 1$, the agent being modeled becomes more and more risk-neutral. Indeed, when $\alpha = 1$, we recover our original model, for expected value is a particular instance of expected utility when the agent is completely risk-neutral.

If we make this simple substitution (in a scenario with two interpretants), we can calculate the utility of a sign, $\mathcal{U}(S)$, as opposed to its value, $\mathcal{V}(S)$, by adjusting equation (2.19) in the following way:

$$\mathcal{U}(S) = \sum_{i} P(S_i) \cdot \big[\max\big(U(I_1/S_i), U(I_2/S_i)\big) - \max\big(U(I_1), U(I_2)\big)\big] . \tag{C.2}$$

Here, $U(I_i)$ and $U(I_i/S_j)$ are simply $V(I_i)$ and $V(I_i/S_j)$, respectively, with their value terms adjusted using equation (C.1):

$$U(I_i/S_j) = \sum_{kl} \big(V(C_k \wedge I_i \wedge O_l) + M\big)^{\alpha} \cdot P(C_k/I_i \wedge O_l) \cdot P(O_l/S_j) , \tag{C.3a}$$

$$U(I_i) = \sum_{kl} \big(V(C_k \wedge I_i \wedge O_l) + M\big)^{\alpha} \cdot P(C_k/I_i \wedge O_l) \cdot P(O_l) . \tag{C.3b}$$

Figure C.1 compares the expected utility of a sign with its expected value (as calculated in section 2.5) for our symptom and sickness scenario. The solid line shows a risk-neutral agent ($\alpha = 1$), for whom the utility of the sign is equivalent to the value of the sign (and thus independent of the value of M) for the parameter values in question.

The non-solid lines, in contrast, show agents with different degrees of risk-adverseness and smaller and larger amounts of money in the bank. As may be seen, as α decreases (and M is held constant), the utility of a sign decreases (reflecting the fact that the agent is more and more risk adverse). In effect, the strategy of acting on their already-known priors, $P(O)$, as opposed to their not-yet-known posteriors, $P(O/S)$, becomes good enough. As may also be seen, the utility of a sign decreases as M increases. This reflects the fact that the outcome of the diagnosis and/or treatment becomes less and less important to the agent as their savings increases. Loosely, the more money they have in the bank, the less a week's salary matters to them.

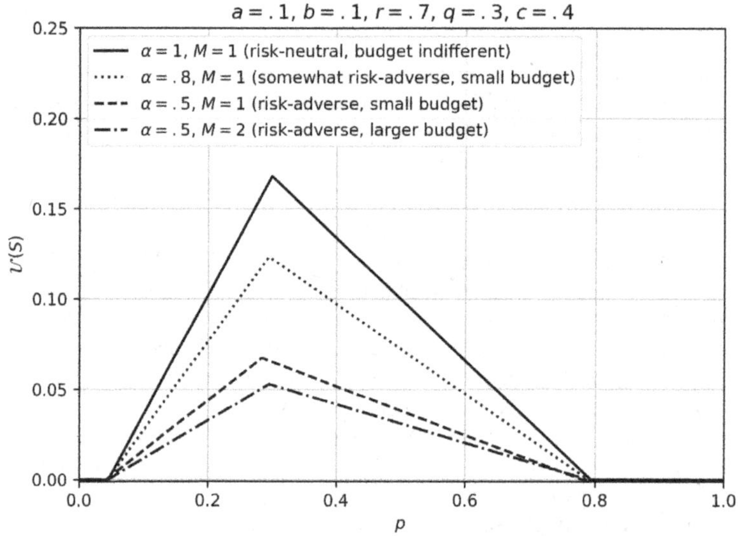

Figure C.1: The Utility of a Sign

To be sure, the magnitude of utility does not tell us anything important in itself (for an infinity of different utility functions can represent the same preferences). More interesting is to calculate the maximum amount of money m that the agent would pay to receive a sign (in the form of a diagnostic), given such a utility function (and hence given particular values of M and α). To find m, we continue to use equation (C.2), along with equation (C.3b), but we change the first term in equation (C.3a) to be

$$\left(V(C_k \wedge I_i \wedge O_l) + M - m\right)^\alpha, \tag{C.4}$$

which reflects the fact that the agent must pay m in all worlds where they choose to have the diagnostic (and *only* those worlds), such that the action they undertake will turn on their posteriors as opposed to their priors. To find the value of m for a given set of parameter values (meaning not only M and α, but also p, a, b, q, r, and c), we need to find the value it has when $\mathcal{U}(S) = 0$, such that

$$\sum_i P(S_i) \cdot \max\left(U(I_1/S_i), U(I_2/S_i)\right) = \max\left(U(I_1), U(I_2)\right). \tag{C.5}$$

This tells us the maximum price m that the agent would be willing to pay for the diagnostic (and hence for the information carried by the sign that the diagnostic returns), given their interpretive ground: any more than this value, and it is not worth getting the diagnostic; any less than this value, and the agent would be willing to pay a little more to get the diagnostic.

Figure C.2 plots m as a function of p for various values of α (holding the other parameter values constant). As may be seen by comparing the dashed and dotted lines with the solid line, which plots $\mathcal{V}(S)$ versus p for the same parameter values, the magnitude of m decreases as an agent becomes more risk-adverse. As may also be seen, as $\alpha \to 1$, $m \to \mathcal{V}(S)$. This is similar to the pattern we saw previously when we graphed $\mathcal{U}(S)$.

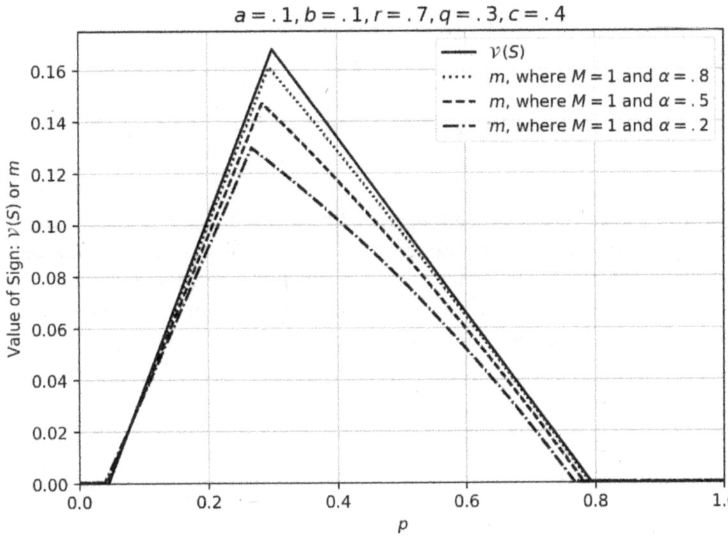

Figure C.2: The Value of a Sign (Revisited)

Figure C.3 shows how all values of m approach one and the same value, itself equal to $\mathcal{V}(S)$, as M increases, regardless of α. In effect, the larger M is, the less important the diagnostic is to the agent (because losing a week's worth of salary becomes less and less important to it). However, and for the same reason, the amount it has to pay for the sign, m, becomes less and less significant, for the cost of the diagnostic is immaterial to an agent with a lot of money in the bank.

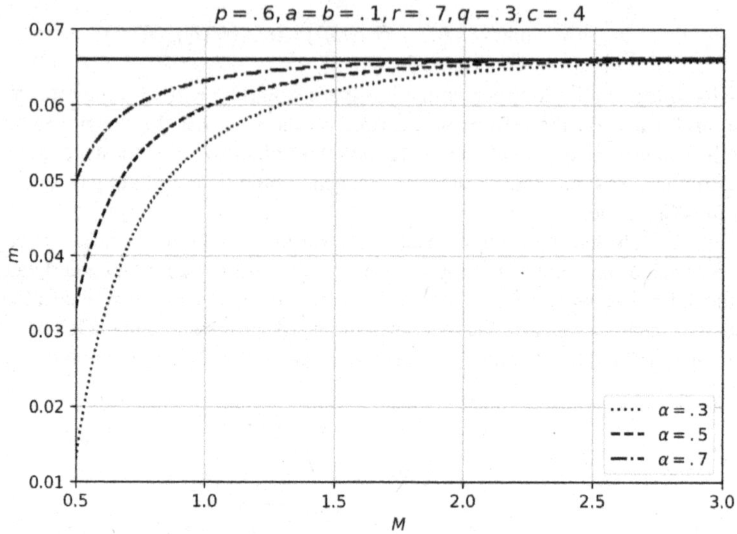

Figure C.3: Convergence of m to $\mathcal{V}(S)$

In short, the two trends proceed together (decreasing utility of loss of income and cost of diagnosis), such that the value of the sign, m, as calculated using our utility function, $U(x+M)^{\alpha}$ approaches the expected value of the sign, $\mathcal{V}(S)$. Note, then, that our original model gave more or less the same result without the explicit introduction of a utility function in the limit of large M and independent of α.

References

Barberis, Nicholas C. 2013. "Thirty Years of Prospect Theory in Economics: A Review and Assessment." *Journal of Economic Perspectives* 27 (1): 173–96.

Barbieri, Marcello. 2008. "Biosemiotics: A New Understanding of Life." *Naturwissenschaften* 95 (7): 577–599.

Bateson, Gregory. 2000. *Steps to an Ecology of Mind: Collected Essays in Anthropology, Psychiatry, Evolution, and Epistemology.* University of Chicago Press.

Bayes, Thomas. 1763. "LII. An Essay towards Solving a Problem in the Doctrine of Chances. By the Late Rev. Mr. Bayes, FRS Communicated by Mr. Price, in a Letter to John Canton, AMFR S." *Philosophical Transactions of the Royal Society of London* 1 (53): 370–418.

Borowski, Audrey. 2024. *Leibniz in His World: The Making of a Savant.* Princeton University Press.

Bourdieu, Pierre. 1977. *Outline of a Theory of Practice.* Cambridge University Press.

Bourdieu, Pierre. 1987. *Distinction: A Social Critique of the Judgement of Taste.* Harvard University Press.

Boyd, Robert, and Peter J. Richerson. 2005. *The Origin and Evolution of Cultures.* Oxford University Press.

Brooks, Daniel R., and Edward O. Wiley. 1988. *Evolution as Entropy.* University of Chicago Press.

Bush, Robert R., and Frederick Mosteller. 1955. *Stochastic Models for Learning.* John Wiley & Sons.

Carnap, Rudolf. 1988. *Meaning and Necessity: A Study in Semantics and Modal Logic.* University of Chicago Press.

Colapietro, Vincent Michael. 1988. *Peirce's Approach to the Self: A Semiotic Perspective on Human Subjectivity.* SUNY Press.

De Finetti, Bruno. 2017. *Theory of Probability: A Critical Introductory Treatment.* John Wiley & Sons.

Deacon, Terrence W. 2011. *Incomplete Nature: How Mind Emerged from Matter.* Norton & Company.

Dretske, Fred I. 1981. *Knowledge and the Flow of Information.* MIT Press.

Enfield, Nicholas James, and Paul Kockelman. 2017. *Distributed Agency.* Oxford University Press.

Favareau, Donald. 2010. *Essential Readings in Biosemiotics: Anthology and Commentary.* Vol. 50. Springer.

Frankfurt, Harry G. 1971. "Freedom of the Will and the Concept of a Person." *The Journal of Philosophy* 68 (1): 5–20.

Goffman, Erving. 1979. "Footing." *Semiotica* 25 (1): 1–30.

Guyer, Jane I. 2004. *Marginal Gains: Monetary Transactions in Atlantic Africa.* University of Chicago Press.

Hacking, Ian. 2001. *An Introduction to Probability and Inductive Logic.* Cambridge University Press.

Hacking, Ian. 2006. *The Emergence of Probability: A Philosophical Study of Early Ideas about Probability, Induction and Statistical Inference.* Cambridge University Press.

Herrnstein, Richard J. 1970. "On the Law of Effect." *Journal of the Experimental Analysis of Behavior* 13 (2): 243–266.

Hitchcock, Christopher. 2021. "Probabilistic Causation." In *The Stanford Encyclopedia of Philosophy,* Spring 2021, edited by Edward N. Zalta. Metaphysics Research Lab, Stanford University.

Hofbauer, Josef, and Simon M. Huttegger. 2008. "Feasibility of Communication in Binary Signaling Games." *Journal of Theoretical Biology* 254 (4): 843–849.

Huttegger, Simon, Brian Skyrms, Pierre Tarres, and Elliott Wagner. 2014. "Some Dynamics of Signaling Games." *Proceedings of the National Academy of Sciences* 111 (3): 10873–10880.

Huttegger, Simon M. 2007. "Evolution and the Explanation of Meaning." *Philosophy of Science* 74 (1): 1–27.

Huygens, Christiaan. 1714. *Christiani Hugenii Libellus de Ratiociniis in Ludo Aleae: Or, The Value of All Chances in Games of Fortune; Cards, Dice, Wagers, Lotteries, &c. Mathematically Demonstrated.* Samuel Keimer.

Jeffreys, Harold. 1998. *The Theory of Probability.* Oxford University Press.

Kahneman, Daniel, and Amos Tversky. 2013. "Prospect Theory: An Analysis of Decision under Risk." In *Handbook of the Fundamentals of Financial Decision Making: Part I,* 99–127. World Scientific.

Kauffman, Stuart. 2003. "Molecular Autonomous Agents." *Philosophical Transactions of the Royal Society of London. Series A: Mathematical, Physical and Engineering Sciences* 361 (1807): 1089–1099.

Kockelman, Paul. 2007a. "Enclosure and Disclosure." *Public Culture* 19 (2): 303–305.

Kockelman, Paul. 2007b. "From Status to Contract Revisited: Value, Temporality, Circulation and Subjectivity." *Anthropological Theory* 7 (2): 151–176.

Kockelman, Paul. 2009. "The Complexity of Discourse." *Journal of Quantitative Linguistics* 16 (1): 1–39.

Kockelman, Paul. 2010. "Enemies, Parasites, and Noise: How to Take Up Residence in a System Without Becoming a Term in It." *Journal of Linguistic Anthropology* 20 (2): 406–421.

Kockelman, Paul. 2011. "Biosemiosis, Technocognition, and Sociogenesis: Selection and Significance in a Multiverse of Sieving and Serendipity." *Current Anthropology* 52 (5): 711–739.

Kockelman, Paul. 2013a. *Agent, Person, Subject, Self: A Theory of Ontology, Interaction, and Infrastructure.* Oxford University Press.

Kockelman, Paul. 2013b. "Information is the Enclosure of Meaning: Cybernetics, Semiotics, and Alternative Theories of Information." *Language & Communication* 33 (2): 115–127.

Kockelman, Paul. 2015. *The Chicken and the Quetzal: Incommensurate Ontologies and Portable Values in Guatemala's Cloud Forest.* Duke University Press.

Kockelman, Paul. 2016. "Meeting the Universe Two-Thirds of the Way (Witchful Thinking)." *Signs and Society* 4 (2): 215–243.

Kockelman, Paul. 2017. *The Art of Interpretation in the Age of Computation.* Oxford University Press.

Kockelman, Paul. 2020a. *Kinds of Value: An Experiment in Modal Anthropology.* Prickly Paradigm Press.

Kockelman, Paul. 2020b. "The Epistemic and Performative Dynamics of Machine Learning Praxis." *Signs and Society* 8 (2): 319–355.

Kockelman, Paul. 2024. *Last Words: Large Language Models and the AI Apocalypse.* Prickly Paradigm Press.

Kratzer, Angelika. 2012. *Modals and Conditionals: New and Revised Perspectives.* Oxford University Press.

Kratzer, Angelika, and Irene Heim. 1998. *Semantics in Generative Grammar.* Wiley-Blackwell.

Kull, Kalevi, Terrence Deacon, Claus Emmeche, Jesper Hoffmeyer, and Frederik Stjernfelt. 2009. "Theses on Biosemiotics: Prolegomena to a Theoretical Biology." *Biological Theory* 4:167–173.

Kullback, Solomon, and Richard A. Leibler. 1951. "On Information and Sufficiency." *Annals of Mathematical Statistics* 22 (1): 79–86.

Laplace, Pierre Simon. 1951. *A Philosophical Essay on Probabilities.* Dover.

Lee, Benjamin. 1997. *Talking Heads: Language, Metalanguage, and the Semiotics of Subjectivity.* Duke University Press.

Lee, Benjamin. 2020. "From Primitives to Derivatives." *Critical Historical Studies* 7 (1): 127–144.

Lee, Benjamin, and Edward LiPuma. 2004. *Financial Derivatives and the Globalization of Risk.* Duke University Press.

Lépinay, Vincent Antonin. 2011. *Codes of Finance.* Princeton University Press.

Levy, Jack S. 1992. "An Introduction to Prospect Theory." *Political Psychology* 13 (2): 171–186.

Lewis, David. 1981. "Causal Decision Theory." *Australasian Journal of Philosophy* 59 (1): 5–30.

Lewis, David. 1986. *On the Plurality of Worlds.* Blackwell.

Lewis, David. 2008. *Convention: A Philosophical Study.* Blackwell.

Lewontin, Richard C. 1983. "The Organism as the Subject and Object of Evolution." *Scientia* 118 (2): 65–82.

Lindley, Dennis V. 1956. "On a Measure of the Information Provided by an Experiment." *Annals of Mathematical Statistics* 27 (4): 986–1005.

MacKenzie, Donald, Fabian Muniesa, and Lucia Siu. 2007. *Do Economists Make Markets? On the Performativity of Economics.* Princeton University Press.

Malinowski, Bronislaw. 2013. *Argonauts of the Western Pacific: An Account of Native Enterprise and Adventure in the Archipelagoes of Melanesian New Guinea.* Routledge.

Maurer, Bill. 2011. *Mutual Life, Limited.* Princeton University Press.

Mayr, Ernst. 1974. "Teleological and Teleonomic, a New Analysis." In *Methodological and Historical Essays in the Natural and Social Sciences,* edited by Robert S. Cohen and Marx W. Wartofsky, 91–117. Springer.

Mayr, Ernst. 1992. "The Idea of Teleology." *Journal of the History of Ideas* 53 (1): 117–135.

McElreath, Richard, and Robert Boyd. 2008. *Mathematical Models of Social Evolution: A Guide for the Perplexed.* University of Chicago Press.

Millikan, Ruth Garrett. 2004. *Varieties of Meaning: The 2002 Jean Nicod Lectures.* MIT Press.

Neapolitan, Richard E. 1990. *Probabilistic Reasoning in Expert Systems: Theory and Algorithms.* Wiley.

Nelson, Philip. 2004. *Biological Physics.* W.H. Freeman & Company.

Nielsen, Michael A. 2015. *Neural Networks and Deep Learning.* Determination Press.

Nowak, Martin A. 2006. *Evolutionary Dynamics: Exploring the Equations of Life.* Harvard University Press.

Oyama, Susan, Russell D. Gray, and Paul E. Griffiths. 2003. *Cycles of Contingency: Developmental Systems and Evolution.* MIT Press.

Parmentier, Richard J. 1994. *Signs in Society: Studies in Semiotic Anthropology.* Indiana University Press.

Pearl, Judea. 1988. *Probabilistic Reasoning in Intelligent Systems: Networks of Plausible Inference.* Morgan Kaufmann.

Pearl, Judea, and Dana Mackenzie. 2018. *The Book of Why: The New Science of Cause and Effect.* Basic Books.

Peebles, Gustav. 2010. "The Anthropology of Credit and Debt." *Annual Review of Anthropology* 39:225–240.

Peirce, Charles S. 1984 [1867]. "On a New List of Categories." In *Writings of Charles S. Peirce: A Chronological Edition, 1867–1871,* edited by Edward C. Moore, 49–58. Indiana University Press.

Peirce, Charles S. 1986 [1878]. "The Doctrine of Chances." In *Writings of Charles S. Peirce: A Chronological Edition, 1872–1878,* edited by Christian J. W. Kloesel, 276–289. Indiana University Press.

Peirce, Charles S. 1986 [1877]. "The Fixation of Belief." In *Writings of Charles S. Peirce: A Chronological Edition, 1872–1878,* edited by Christian J. W. Kloesel, 242–257. Indiana University Press.

Polanyi, Karl. 2001. *The Great Transformation: The Political and Economic Origins of Our Time.* Beacon Press.

Ramsey, Frank P. 2016. "Truth and Probability." In *Readings in Formal Epistemology,* edited by Horacio Arló-Costa, Vincent F. Hendricks, and Johan Van Benthem, 21–45. Springer.

Reichenbach, Hans. 1956. *The Direction of Time.* University of California Press.

Reif, F. 1965. *Fundamentals of Statistical and Thermal Physics.* McGraw Hill.

Roth, Alvin E., and Ido Erev. 1995. "Learning in Extensive-Form Games: Experimental Data and Simple Dynamic Models in the Intermediate Term." *Games and Economic Behavior* 8 (1): 164–212.

Rumelhart, David E., Geoffrey E. Hinton, and Ronald J. Williams. 1986. "Learning Representations by Back-Propagating Errors." *Nature* 323 (6088): 533–536.

Russell, Stuart, and Peter Norvig. 2002. *Artificial Intelligence: A Modern Approach.* Pearson.

Sahlins, Marshall. 2013. *Stone Age Economics.* Routledge.

Savage, Leonard J. 1972. *The Foundations of Statistics.* Courier Corporation.

Shannon, Claude E. 1949. "Communication Theory of Secrecy Systems." *The Bell System Technical Journal* 28 (4): 656–715.

Shannon, Claude E. 1948. "A Mathematical Theory of Communication." *The Bell System Technical Journal* 27 (3): 379–423.

Skyrms, Brian. 1982. "Causal Decision Theory." *The Journal of Philosophy* 79 (11): 695–711.

Skyrms, Brian. 2010. *Signals: Evolution, Learning, and Information.* Oxford University Press.

Smith, John Maynard. 1982. *Evolution and the Theory of Games.* Cambridge University Press.

Smith, John Maynard, and David Harper. 2003. *Animal Signals.* Oxford University Press.

Smith, John Maynard, and George R. Price. 1973. "The Logic of Animal Conflict." *Nature* 246 (5427): 15–18.

Stegmann, Ulrich E. 2005. "John Maynard Smith's Notion of Animal Signals." *Biology & Philosophy* 20 (5): 1011–1025.

Strogatz, Steven H. 2018. *Nonlinear Dynamics and Chaos: With Applications to Physics, Biology, Chemistry, and Engineering.* CRC Press.

Taylor, Charles. 1992. *Sources of the Self: The Making of the Modern Identity.* Harvard University Press.

Thorndike, Edward L. 1898. "Animal Intelligence: An Experimental Study of the Associative Processes in Animals." *Psychological Review: Monograph Supplements* 2 (4): 101–109.

Tomlinson, Gary. 2023. *The Machines of Evolution and the Scope of Meaning.* MIT Press.

Uexküll, Jakob von. 2013. *A Foray into the Worlds of Animals and Humans: With a Theory of Meaning.* University of Minnesota Press.

Urban, Greg. 2001. *Metaculture: How Culture Moves through the World.* University of Minnesota Press.

Urban, Greg. 2010. "A Method for Measuring the Motion of Culture." *American Anthropologist* 112 (1): 122–139.

Veblen, Thorstein. 2017. *The Theory of the Leisure Class.* Routledge.

Von Fintel, Kai, and Irene Heim. 2011. "Intensional Semantics." *Unpublished Lecture Notes.*

Weatherby, Leif. 2016. *Transplanting the Metaphysical Organ: German Romanticism between Leibniz and Marx.* Fordham University Press.

Weber, Max. 2017. *Methodology of Social Sciences.* Routledge.

Weber, Max. 2019. *Economy and Society.* Harvard University Press.

Weirich, Paul. 2020. "Causal Decision Theory." In *The Stanford Encyclopedia of Philosophy,* Winter 2020, edited by Edward N. Zalta. Metaphysics Research Lab, Stanford University.

Whitehead, Alfred North. 1920. *The Concept of Nature.*

Wright, Sewall. 1921. "Correlation and Causation." *Journal of Agricultural Research* 20 (7): 557–585.

Wright, Sewall. 1934. "The Method of Path Coefficients." *Annals of Mathematical Statistics* 5 (3): 161–215.

Index